Microwave Propagation and Remote Sensing

Atmospheric Influences with Models and Applications

T0225393

Microwave Propagation and Remote Sensing

Atmospheric Influences with Models and Applications

Pranab Kumar Karmakar

CRC Press
Taylor & Francis Group
Boca Raton   London   New York

CRC Press is an imprint of the
Taylor & Francis Group, an **informa** business

CRC Press
Taylor & Francis Group
6000 Broken Sound Parkway NW, Suite 300
Boca Raton, FL 33487-2742

First issued in paperback 2017

© 2012 by Taylor & Francis Group, LLC
CRC Press is an imprint of Taylor & Francis Group, an Informa business

No claim to original U.S. Government works
Version Date: 20110708

ISBN 13: 978-1-138-07643-3 (pbk)
ISBN 13: 978-1-4398-4899-9 (hbk)

This book contains information obtained from authentic and highly regarded sources. Reasonable efforts have been made to publish reliable data and information, but the author and publisher cannot assume responsibility for the validity of all materials or the consequences of their use. The authors and publishers have attempted to trace the copyright holders of all material reproduced in this publication and apologize to copyright holders if permission to publish in this form has not been obtained. If any copyright material has not been acknowledged please write and let us know so we may rectify in any future reprint.

Except as permitted under U.S. Copyright Law, no part of this book may be reprinted, reproduced, transmitted, or utilized in any form by any electronic, mechanical, or other means, now known or hereafter invented, including photocopying, microfilming, and recording, or in any information storage or retrieval system, without written permission from the publishers.

For permission to photocopy or use material electronically from this work, please access www.copyright.com (http://www.copyright.com/) or contact the Copyright Clearance Center, Inc. (CCC), 222 Rosewood Drive, Danvers, MA 01923, 978-750-8400. CCC is a not-for-profit organization that provides licenses and registration for a variety of users. For organizations that have been granted a photocopy license by the CCC, a separate system of payment has been arranged.

Trademark Notice: Product or corporate names may be trademarks or registered trademarks, and are used only for identification and explanation without intent to infringe.

Visit the Taylor & Francis Web site at
http://www.taylorandfrancis.com

and the CRC Press Web site at
http://www.crcpress.com

Dedicated to the Lotus feet of Goddess Shubhankari

Contents

Preface

Radio wave is a term arbitrarily applied to electromagnetic waves in the frequency range 0.001 to 10^{16} Hz. The lower limit of radio waves propagated in free space is 3×10^{11} m and the upper limit is 3×10^{-8} m. It is relevant to note that these limits of the radio wave spectrum are considerably wider than they were assumed to be until quite recently. In the low frequency side, for example, 0.001 Hertz (H_2) is an arbitrary limit, for as science advances one will have to deal with the progressively lower frequency side. But, on the high frequency side the radio wave spectrum limited to 10^{12} Hz was called submillimeter wave. However, from the engineering point of view, the wave of frequency ranging from 3 to 300 GHz may sometimes be called microwaves. My intention has been to present an elementary and short account of representative aspects of microwave propagation through the atmosphere and sensing the atmosphere by deploying microwaves.

This text will be concerned with the unguided propagation of microwaves in the neutral atmosphere. This finds many extremely important applications in science and engineering. These include transmission of intelligence and radiometric application for probing the atmosphere. The radiometric application in the microwave band for characterization and modeling of the atmospheric constituents is also included in this text, which we call remote sensing of the neutral atmosphere. It demands, therefore, an updated account of the results in this field of research to be gathered and presented in a single volume with an elaborated discussion in microwave propagation through the neutral atmosphere. This book is intended to deal with different aspects of microwave propagation like reflection and refraction in great detail with special emphasis on microwave absorption by different atmospheric constituents including rain. In all chapters, the literature references are intended only to be representative of some of the reviews and classic papers along with enough of a selection from recent work to enable the reader to get an impression of the nature of the current activity. The selection of citations from such a wealth of excellent work is more random than calculated. I extend my apologies for those omitted through the action of laws of chance.

In fact, radio wave propagation above 10 GHz is highly influenced by the prevalent atmospheric condition, in particular composition, and characteristic of the troposphere. In order to have a deeper insight about the microwave propagation, one has to be thorough enough regarding the troposphere. In this context, some emphasis has been given to exemplify the characteristic phenomena of the troposphere. The most important key parameters of the troposphere are water vapor and rain so long as the microwave propagation is concerned. In the first chapter of this text, the derived parameters of water vapor, namely, vapor pressure, density, ray bending, and types of rain and its distribution pertaining to microwave propagation, are dealt with in great detail.

The second chapter deals with free space propagation. In the case of microwave communication, free space field due to directional transmitting aerial and the power to be received at the receiving terminal along with free space loss calculations have been considered. When radio wave propagates through neutral atmosphere, which does not have any free charges, it induces some displacement current in suspended particles of the atmosphere. This causes absorption of energy, which is manifested as heat as a result of which the radio wave gets attenuated. This is discussed in detail in light of Maxwell's equation.

The third chapter deals with reflection, interference, and polarization of the electromagnetic wave while it is propagating through the atmosphere. Emphasis has been given to consider these effects due to curvature of the Earth's surface. Along with this, a discussion has been made about the mechanism of ground wave propagation.

Chapter 4 addresses radio refraction, which ultimately produces delay. A detailed discussion has been made to highlight the dependence of frequency on radio refractivity. The atmospheric turbulence has also been addressed. The ultimate effect of turbulence is to produce scintillation in the microwave band. The modeling method is discussed in this chapter. Another important factor in this band is tropospheric ducting. The physics behind atmospheric ducting is also addressed.

The radio wave while propagating through atmosphere gets attenuated due to the presence of suspended atmospheric particles. This absorption amounts to different values depending on frequency of operation. A detailed discussion is included about the methodology of modeling of water vapor attenuation by using the radiosonde data. All these are included in Chapter 5.

It is known that most of the water in air remains as water vapor rather than as liquid or solid hydrometeors. It is understood that if the ambient temperature at a certain height is 0°C or below, the hydrometeors are ice. Below this height where temperature is 0°C, ice starts melting and transfers into rain. But the rising air currents prevent the fall of rain. As water condenses, it forms ice crystals that are small enough to be supported by air currents. These particles are clubbed together until they become too heavy for rising currents to support. These heavy particles fall as rain. However, rain is not uniform but can be approximated to a group of rain cells in the form of a cylinder of uniform rain, for simplicity's sake, extending from the cloud base height to the ground. Rain produces a significant attenuation on radio waves of frequencies beyond 10 GHz. When the rain cells intercept the propagation path, attenuation is caused. Each water droplet may be considered an imperfect conductor. The incoming radio wave induces displacement current. As the dielectric constant of water is 80 times larger than that of air, the density of this displacement current is large. On the other hand, the density of the displacement current is proportional to frequency. This implies that the displacement current will be larger in the microwave or millimeter wave band. This is discussed in detail in Chapter 6. In addition to this absorption, the other important phenomenon is the rain scattering. In fact, scattering depends on the particle size and the frequency concerned. Thus the absorption loss and the scattering loss, when summed up, produce attenuation. The prediction of rain attenuation generally starts from known point rainfall rate statistics, considering the vertical and horizontal structures of rain cells. Having the knowledge of rain structures and by using the climatological parameters, one can estimate rain attenuation. Attenuation can also be estimated from the radiometric measurements of sky-noise temperature with certain assumptions made regarding the atmospheric temperature. The vertical extent of rain can also be estimated from the meteorological measurements of the height of $0°$ isotherm and the radar reflectivity measurement from which rain attenuation along the vertical path can also be estimated. All these are discussed in Chapter 6.

Chapter 7 is devoted to attenuation of microwaves other than rain, like snow, fog, ambient water vapor, and nonprecipitable liquid water. Radiometric methods of measurement of water vapor and hence attenuation is discussed. The presence of cloud influences radiometric attenuation. To surmount this difficulty, dual frequency/multifrequency methodology is discussed. Size distribution of aerosols by using optical wavelengths is also discussed.

It is my highest aim, objective, and endeavor to concentrate on the basic philosophy and to balance between dexterity of treatment and the depth of the subject matter. Primary emphasis has been laid on the physical principles and to arouse the power and ability of the graduate and undergraduate audience in one particular area of microwaves, namely, propagation and ground-based remote sensing. It will be worthwhile to mention that at the time of preparation of the manuscript, I freely consulted the existing textbooks by eminent authors. Had I not received the encouragement, support, and cooperation from a lot of people around me, my dream would have remained a dream alone. I am grateful to all of them.

It will be unjust for me if I do not mention my wife, Anupama Karmakar, not only for her encouragement but also for the hazards that she underwent for the last couple of years. Otherwise it would not have been possible for me to work with rapt attention. Last, but not the least, I must mention my son, Anirban Karmakar, who has extended his untiring effort in drawing the figures cited in the chapters.

With all my modesty, I wish to say that the publication of such a book covering a syllabus of propagation and remote sensing in a single volume is a challenging task. The support extended by the people of different categories of CRC Press, Taylor & Francis are noteworthy while developing this present text in its final form. Hence, the heartiest thanks to CRC Press.

Pranab Kumar Karmakar
Institute of Radiophysics and Electronics
University of Calcutta
Kolkata, India

Author

Pranab Kumar Karmakar obtained his MSc in physics in 1979 and PhD in the area of microwave propagation and remote sensing in 1990 from the University of Calcutta, India. Associated with the Department of Radiophysics and Electronics at Calcutta University since 1988, he is involved in both teaching and research work. He has more than forty-five publications in national and international journals of repute. Karmakar also has more than thirty conference articles to his credit. He was awarded the Young Scientist Award of URSI (International Union of Radio Science) in 1990. He has been a visiting scientist at the Remote Sensing Lab, University of Kansas (USA); Centre for Space Science, China; and Satellite Division, National Institute for Space Research (INPE), Brazil. He has also been awarded the South-South fellowship of TWAS (The Academy of Sciences for the Developing World) in 1997. A book entitled *Outlines of Physics* (Volumes I and II) published by Platinum Publisher, Kolkata, India, also goes to the credit of P.K. Karmakar.

Karmakar's current area of research includes microwave/millimeter wave propagation, microwave remote sensing, and atmospheric modeling.

1

Outlines of Radio Waves and Troposphere

1.1 General Perspective

Thermal radiation is the process of transmission of energy from a source either by motion of material particles or by waves as in sound, light, and x-rays. Here, the rays are the curved lines along which the energy is sent from the source. One can experience radiation in daily life; when he holds his hand in front of a hot object and radiant heat is felt. We can feel radiant heat as we stay out in the sun. Moreover we need light to see the surroundings. All these are the examples of electromagnetic radiation. According to the wavelength or frequency, this radiation is grouped or classified and named as x-rays, ultraviolet rays, solar rays, infrared, and radio waves. But, we will restrict ourselves only with radio waves including microwaves and millimeter waves. Table 1.1 is presented for a better understanding of grouping or classifications or naming of electromagnetic waves. The movement or oscillation of electrical charges in a conductive material or in antenna produces the electromagnetic waves. These waves are radiated away from the transmitting antenna and are then intercepted by a receiving antenna such as a TV antenna or hand-held device (e.g., cellular phone). Electromagnetic waves travel through space at the speed of light. The wavelength and frequency of an electromagnetic wave are inversely related as

$$\lambda = \frac{c}{f}$$

where c is the speed of light, f is the frequency of electromagnetic wave, and λ is the wavelength.

The term *microwave* is used as a generic name to include the centimeter, millimeter, and submillimeter region of the spectrum. Usually the frequency band extending from 3 to 300 GHz is named as the microwave band. But, in

TABLE 1.1

Classification of Electromagnetic Waves According to Wavelength

Frequency	Wavelength	Photon Energy	Band Designations	Generic Name
3×10^{20}	1 pm	1.2 Mev		γ-ray
3×10^{19}	10 pm	120 kev		Medical x-ray
3×10^{18}	1 A	12 kev		Medical x-ray
3×10^{17}	1 nm	1.2 kev		x-ray
3×10^{16}	10 nm	120 ev		
3×10^{15}	100 nm	12 ev		Ultraviolet
3×10^{14}	1 μm	1.2 ev		Visible
3×10^{13}	10 μm	0.12 ev		Near infrared
3×10^{12}	100 μm	1.2×10^{-2} ev		Far infrared
3×10^{11}	1 mm	1.2×10^{-1} ev		Far infrared
3×10^{10}	1 cm	1.2×10^{-4} ev		
3×10^{9}	10 cm	1.2×10^{-5} ev		Microwave
3×10^{8}	1 m	1.2×10^{-6} ev	UHF	
3×10^{7}	10 m	1.2×10^{-7} ev	VHF	
3×10^{6}	100 m	1.2×10^{-8} ev	HF	
3×10^{5}	1 km	1.2×10^{-9} ev	MF	
3×10^{4}	10 km	1.2×10^{-10} ev	LF	Radio wave
3×10^{3}	100 km	1.2×10^{-11} ev		
3×10^{2}	1 Mm	1.2×10^{-12} ev	Audio	
3×10^{1}	1 Mm	1.2×10^{-13} ev	Audio	
3×10^{0}	100 Mm	1.2×10^{-14} ev		

fact, manufacturers have subdivided the millimeter wave band (30–300 GHz) into the following overlapping frequencies:

18–26.5 GHz K band
26.5–40 GHz Ka band
33–50   GHz Q band
40–60   GHz U band
50–75   GHz V band
60–90   GHz E band
75–110   GHz W band
90–140   GHz F band
110–170 GHz D band
140–220 GHz G band

Submillimeter waves extending from 300 to 3000 GHz are similar in behavior as those of the millimeter waves. They do not differ substantially from millimeter wave particularly when the propagation effects are considered. The millimeter and submillimeter waves lie between microwave and infrared. But still, in this text we shall call all these bands microwaves. This text will be concerned with the unguided propagation of microwaves in the terrestrial atmosphere. In other words, we shall leave out the propagation of radio waves directed by man-made structures, such as coaxial lines, waveguides, and strips.

Unguided propagation is most often applied in the fields of science and technology. These include the transmission of intelligence over short and long distances in telephony, television, and the detection and ranging of objects in radar; navigation use of radio locating or direction of aircraft and distance measurements by radio means. One of the most important applications of microwave technology is in remote sensing to probe the atmosphere. In unguided propagation of radio waves we use the transmission media as the neutral media, such as troposphere around the earth.

1.2 Troposphere

1.2.1 Composition and Characteristics

The earth's atmosphere is a collection of many gases along with suspended particles in the form of solids and liquids. Among the gases present, nitrogen and oxygen occupy about 99% of the total volume. The variable components in the atmosphere are water vapor, ozone, sulfur dioxide, and dust. These are called minor constituents of the atmosphere. But argon and carbon dioxide are also considered to be the abundant gases.

The lower most portion of the earth's atmosphere extending from the surface up to a height of 8 to 10 km at polar region and 10 to 12 km at the moderate latitudes and 16 to 18 km at the equator is called troposphere. The percentage of gas components in the atmosphere is presented in Table 1.2.

The gas components shown in Table 1.2 for the most part do not vary with height. But there is a strong exception in case of water vapor. It is strongly dependent on weather conditions and decreases with height obeying the relationship

$$\rho(h) = \rho_0 \exp\left(-\frac{h}{b}\right) \qquad (1.1)$$

where, ρ is the water vapor density at height h and ρ_0 is the surface water vapor density, and h is the scale height where the water vapor density becomes, $\frac{1}{e}$ times the surface value.

The most important property of the troposphere is the temperature. The average vertical temperature gradient of the troposphere is 6–7°C/km. The

TABLE 1.2

Constituents of Dry Atmosphere at Sea Level

Gas	Molecular Weight	Partial Pressure (mb)	Molecules per Cubic Centimeter	Percent by Volume	Percent by Weight
Dry air	29	1012.00	2.7×10^{19}	100.00	100.00
N_2	28	791.00	2.11×10^{19}	78.09	75.53
O_2	32	212.00	5.6×10^{18}	20.95	23.14
Ar	39	9.45	2.5×10^{17}	0.93	1.28
CO_2	44	0.31	8.1×10^{15}	0.03	0.046
Ne	20	1.2×10^{-2}	4.9×10^{14}	1.8×10^{-3}	1.25×10^{-3}
H_2	2	5.1×10^{-4}	1.3×10^{13}	5×10^{-5}	3.48×10^{-6}
He	4	5.3×10^{-3}	1.4×10^{14}	5.24×10^{-4}	7.24×10^{-5}

annual average temperature in the polar region is −55°C and 80°C at the equator. Troposphere is characterized by temperature decrease with height. The height at which the temperature ceases to decrease is known as tropopause. The reason behind this decrease in temperature lies in the fact that the bulk of solar energy after incidence on the earth's surface gets heated, which in turn, radiates heat in the upward direction. Consequently, air layers adjacent to the ground rise in temperature and go upward toward a comparatively colder air, which is in turn heated and moves upward, and so on. This nonuniform heating of the layers of the atmosphere produce the convection current, which ultimately decides the temperature distribution in the atmosphere.

The most important key characteristic of the troposphere is the water vapor concentration, so far as the tropospheric propagation is concerned, along with corresponding temperature and water vapor pressure.

The water vapor density $\rho(g/m^3)$ may be deduced by using the relationship

$$\rho = \frac{e \times 18 \times 10^2}{8.31 \times T_D} \, g/m^3 \tag{1.2}$$

where T_D is the dew point temperature and e is the water vapor pressure, which is discussed in Section 1.2.2.

The monthly mean profiles of water vapor density over Kolkata, West Bengal, India, (22° N) shows that the water vapor density bears a maximum 25 g/m³ during the month of July. However, according to National Weather Service date bank (United States), assigned to NASA Wallops Island Facility during August 14–25, 1975, reveals that maximum water vapor density is 11 g/m³ over Haystack Laboratory in Westford, Massachusetts. To get more detailed information about the water vapor distribution over the whole of the Indian zones, we divide India into two categories from the meteorological point of view (Sen and Karmakar 1988):

Category 1
    a. Indian Island
    b. South East Coast
    c. South West Coast
    d. East Coast
    e. West Coast

Category 2
    a. Northern plane
    b. Central plane
    c. Western plane
    d. Southern plane
    e. Desert area
    f. Assam valley

"The Atlas of Tropospheric Water Vapor over the Indian Subcontinent" is published by National Physical Laboratory, New Delhi. This is compiled from the data collected by India Meteorological Department on a routine basis over sixteen stations from 1968 to 1971, through 1000 mb and 50 mb pressure level of the atmosphere. These data were used to find the integrated water vapor content, defined as water vapor content in a cylinder of unit square meter base and 10 km height. It is expressed in grams per square meter (g/m^2) and can be found out by using the relation

$$W = \int_0^h \rho(h)dh \qquad (1.3)$$

The seasonal variation of the integrated water vapor content at 0000 GMT (0530 1ST) and 2230 GMT (1730 1ST) over the coastal region are shown in Figure 1.1, respectively, whereas the distributions for planes are presented in Figure 1.2.

1.2.2 Relationships for Determination of Vapor Pressure

During 19th century, the measurements of saturation vapor pressure (SVP) draw a special attention to the scientist. The developments were taking place with the advent of the physics of heat and thermodynamics, which

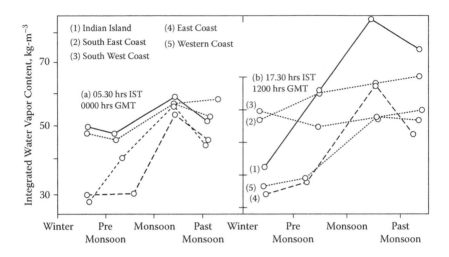

FIGURE 1.1
Variation of integrated water vapor content over the coastal region of India.

culminated in the establishment of the second law of thermodynamics. During this time Clapeyron (1834) found a relationship between saturation vapor pressure (e), and temperature (T(K)) as (Gibbins 1990)

$$\frac{de}{dT} = \frac{L}{T(v_g - v_f)} \tag{1.4}$$

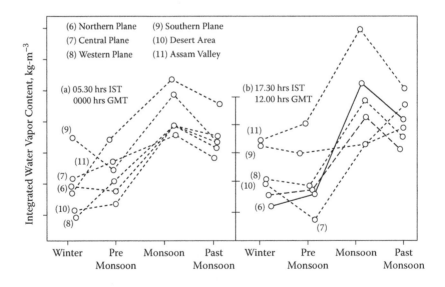

FIGURE 1.2
Variation of integrated water vapor content over planes and valleys of India.

where L is the latent heat of vaporization; v_g is the specific volume of vapor, and v_f is the specific volume of liquid. This relation was proved thermodynamically by Clausius (1850) and is nowadays universally accepted as the Clausius–Clapeyron equation. But, while finding the SVP, Clausius assumed that water vapor behaves as a perfect gas and latent heat of vaporization is constant, that is independent of temperature. To get rid of these ideas, Goff and Gratch (1946) considered that water vapor does not behave as an ideal gas and wrote the equation of state in a quasi-virial form as

$$ev_g = RT - Be - C^2 \tag{1.5}$$

where B and C are the second and third virial coefficients. Now, using empirical temperature dependent relationships for the virial coefficients, Goff and Gratch obtained the following equation over plane surface of pure water for temperature range –50°C to 102°C as (Gibbins 1990)

$$log_{10}e_w = -7.90928\left\{\frac{T_s}{T}-1\right\} + 5.02808log_{10}\left\{\frac{T_s}{T}\right\} - 1.3816 \times 10^{-7}\left\{10^{11.344(1-\frac{T}{T_s})}\right\}$$

$$+ 8.1328 \times 10^{-3}\left\{10^{-3.49149(\frac{T_s}{T}-1)} - 1\right\} + log_{10}e_{ws} \tag{1.6}$$

where T_S is the steam point temperature defined as 373.16 K and e_{wS} is the saturation vapor pressure of pure liquid water at steam point temperature and it is 101.3246 hpa. But after some simple rearrangement of Equation (1.6) it can be rewritten (Queney 1974) as

$$log_{10}e_w = 23.8319 - \frac{2948.964}{T} - 5.02808log_{10}T - 29810.16\exp\left(-0.0699382T\right)$$

$$+ 25.21935\exp\left(-\frac{2999.924}{T}\right) \tag{1.7}$$

However, assuming temperature dependent latent heat of vaporization, Henderson–Sellers (1984) arrived at the following equation:

$$e_w = 6.1078 \times \exp\left\{6828.6\left[\frac{1}{273}-\frac{1}{T}\right]5.1701\ln\frac{T}{273}\right\} \tag{1.8}$$

Here, 0°C is clearly identified with 273 K.

We have so far considered the vapor as a pure vapor over its liquid state. But while applying the equations for finding water vapor pressure in the earth's atmosphere, SVP needs to be corrected. The earth's atmosphere, however, neglecting any impurities or pollutants, is just a mixture comprised of dry air and water vapor of which the latter amounts to little more than 1% at sea level. Thus for moist air, the effect is to increase SVP by up to 5% depending on temperature and pressure.

This increase is characterized by the so-called enhancement factor f_W (Hyland 1975). This is given by Buck (1981) for higher accuracy

$$f_w = 1.00072 + \{3.20 \times 10^{-6} + 5.9 \times 10^{-10} t^2\} P \tag{1.9}$$

where P is the barometric pressure.

1.3 The Effective Earth's Radius

Microwave radio communication, for obvious reasons, requires a clear path between the transmitting and receiving antenna. This clear path is usually known as line of sight (LOS). In this path we should not expect any obstruction. This radio LOS, called radio visibility, differs from optical LOS, called optical visibility. Radio LOS takes into account the concepts of Fresnel ellipsoids and their clearance criteria as shown in Figure 1.3. But due to atmospheric refraction, the path of direct and reflected rays are curved toward the earth (Figure 1.4). This figure shows the ray path (broken lines) without refraction. Hence at the receiving terminal, the received radio signal after interference will not produce the same value for their curved path as expected. Second, propagation velocity in the lower troposphere will be less than it is in the upper troposphere.

To surmount this difficulty, Schelling et al. (1993) assume an earth approximately larger than the actual earth so that the curvature of radio ray may be incorporated in the curvature of the effective earth (Dolukhanov 1971), thus leaving the relative curvature of the two as the same and allowing the

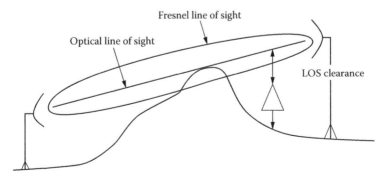

FIGURE 1.3
Schematic diagram of optical and radio link.

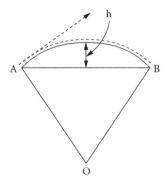

FIGURE 1.4
Bending of radio waves around the earth's surface.

radio rays to be drawn as a straight line over the effective earth rather than as curved lines over the true earth (Figure 1.5). From this figure we write,

$$\frac{1}{a} - \frac{1}{R} = \frac{1}{a'} - \frac{1}{\infty}$$

or, where $\frac{1}{a} - \frac{1}{R}$ is defined as the relative curvature.

$$a' = \frac{a}{1 - \frac{a}{R}} \tag{1.10}$$

where a is the true radius of the earth and a' is the effective earth's radius.

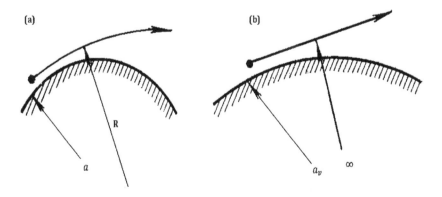

Ray path: (a) over the true earth
(b) over the equivalent earth

FIGURE 1.5
Ray path over (a) the true earth (b) the equivalent earth.

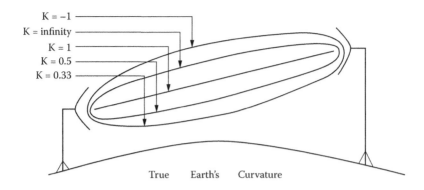

FIGURE 1.6
Effective earth's radius. K = true earth's radius.

Hence, the effective earth's radius factor, K, is defined as the factor that is multiplied by the actual earth's radius a to give the effective earth's radius. Thus $k = a'/a$. Due to earth's curvature and refraction of radio signal, each site must have a minimum elevation with respect to antenna height. The effects of refraction are significant within an area round the direct path known as the Fresnel zone (Figure 1.6). We usually use 60% of the full depth of Fresnel zone for radio LOS.

The value of k can be calculated for a given area based on refractivity gradient available from local chart. For standard atmosphere, $k = 1.33 = 4/3$. Higher values of k would mean a greater amount of bending of radio waves toward the earth's surface and consequently would result in extension of radio visibility. The value of radio horizon distance for a particular value of k is given by $\sqrt{2kah}$ where h is the antenna height.

1.4 Radio Link

Radio link or circuit is essentially a transmitter, receiver, and the propagation medium. We shall consider only radio transmission through natural medium such as atmosphere around the earth.

Radio engineers usually deal with three types of radio circuits. First, the transmitter and receiver both are stationed on the surface of the earth. Hence, in this case, the transmitted signal reaches the receiver by ionospheric reflection (Figure 1.7). Second, the radio waves can reach the receiver through diffraction, scattering, or line of sight (Figure 1.8). The other type of link is between an earth-based station and space probe.

In Figure 1.7, the transmitted signal cannot reach the receiver directly. The transmitted signal illuminates the man-made or natural body at C. Then

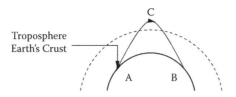

FIGURE 1.7
A simple radio link: A denotes transmitter, B denotes the receiver.

from C, the signal is picked up at after reflection or scattering. In some cases, a scatter system may have its transmitter and receiver collocated. An example is radar. It is to be noted that the scatterer may lie within the troposphere. In that case it is named the troposcatter link (Figure 1.8).

1.5 Classification of Radio Waves According to Propagation Mechanism

1.5.1 Direct Wave

The simplest type of radio propagation is the free-space propagation. By the term *free space* we mean the medium is isotropic, homogeneous, and loss free. In this case we assume that the radio wave propagates along a straight line. In other words, the transmitted wave reaches the receiver through a straight line and hence the name is direct wave (Figure 1.9). So, a direct wave is a radio wave that propagates through an isotropic, homogeneous, and loss-free medium along straight lines. That is to say, a direct wave propagates through free space. Thus it is not influenced by a magnetic field or gravitational field. So, free space is unlikely to exist anywhere. But from some practical point of view, we sometimes talk about free space propagation to get a better insight

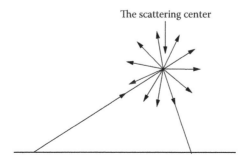

FIGURE 1.8
Schematic diagram of a simple scatter radio link.

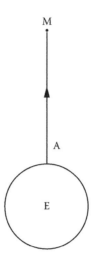

FIGURE 1.9
Propagation of radio wave.

about propagation conditions, especially in case of microwave propagation. It may also be noted that in free space, the power density of the transmitted signal obeys the inverse square law, that is, power density (P) is given by

$$P = \frac{P_t}{4\pi r^2} \tag{1.11}$$

where P_t is the transmitted power and r is the distance from the source where the power density is observed.

We know, the characteristic impedance of a medium is given by

$$z = \sqrt{\frac{\mu}{\epsilon}} \tag{1.12}$$

where, ϵ is the permittivity of the medium and μ is the permeability of the medium.

Now, for free space,

$$\mu = 4\pi \times 10^{-7} = 1.257 \times 10^{-6} \text{ H/m}$$

$$\epsilon = \frac{1}{36\pi} \times 10^{-9} = 8.854 \times 10^{-12} \text{F/m}$$

So, on substitution of these values, we get the impedance

$$z = \sqrt{\frac{4\pi \times 10^{-7}}{\dfrac{1 \times 10^{-9}}{36\pi}}} = 120\pi = 377\Omega$$

Let us now, calculate the field intensity of electromagnetic wave, which is being propagated through free space. We know that the power density, P, and field intensity is related through the characteristic impedance, Z, as

$$P = \frac{E_{rms}^2}{Z}$$

$$E_{rms}^2 = PZ = \frac{P_t}{4\pi r^2} \times 120\pi = 30 \frac{P_t}{r^2} \tag{1.13}$$

$$E_{rms} = \sqrt{30P_t/r} \ V/m$$

Thus field intensity is inversely proportional to distance from source.

1.5.2 Ground Wave or Surface Wave

When radio wave propagates close to the ground and follows the curvature of the globe due to diffraction it is called a ground wave. Diffraction of a radio wave occurs when the wave is propagated across an obstacle. The earth may be regarded as a spherical obstacle and radio waves are diffracted around curvature of the globe. Usually long waves (30–300 MHz) suffer diffraction. Considering earth's surface is partially conducting, the radio waves become attenuated and wave forms also get distorted. On the other hand, the spherical form of the earth causes the radio waves propagated along its surface to diffract. Moreover, since the troposphere is an inhomogeneous medium with time varying properties due to its varying weather conditions, its refractive index gradually decreases with height. This refraction from the local irregularities in the troposphere causes the bending of the radio path in such a way so that they can also follow the curvature of the earth. By this way troposphere affects the propagation of ground waves resulting in promotion of communication over long distances.

Local inhomogeneous substances scatter the radio waves (Figure 1.10) and manifest itself only of wavelengths shorter than 10 m along with slight diffraction. Sometimes, sudden changes in irregularities due to sudden change in weather conditions may cause a steeper decrease of refractive index in the lower troposphere. This results in the formation of duct and consequently the radio wave gets trapped in that duct. This happens, usually, to waves shorter than 3 m and move along the surface of the earth.

1.5.3 Tropospheric Wave

We know that an electromagnetic wave front spreads uniformly in all directions. If the particular point on a wave front was followed over time, the collection of point positions would define a ray. This ray path may be considered as straight lines except where the presence of the earth and its atmosphere

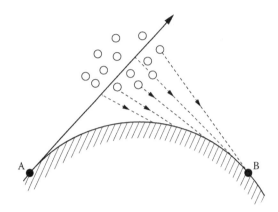

FIGURE 1.10
Scattering from irregularities in the troposphere.

tend to alter the path. Thus except in some unusual circumstances, frequencies above 30 MHz or shorter than 3 m generally travel in straight lines. This part of the electromagnetic wave is called tropospheric waves as they travel through the troposphere.

Sometimes the ground waves may be considered a tropospheric wave depending upon the nature of propagation. To be clearer about this, usually when the wave strictly follows the propagation around the curvature of the earth, the wave is a ground wave although it propagates through the troposphere. Sometimes the ground wave propagation refers to beyond-the-horizon propagation. An example is radio broadcasting in daytime.

1.6 Radio Refractivity and Delay through the Atmosphere

Out of all the basic parameters for tropospheric radio wave propagation, refraction needs some special attention and hence delay. Group delay introduced by tropospheric water vapor content, for radio signal in very high frequency (VHF; 30–300 MHz), ultrahigh frequency (UHF; 300–3000 MHz), and microwaves and millimeter waves, may be comparable and even greater than ionospheric delay.

The troposphere may be treated as a mixture of dry air and water vapor. The refractive index of this mixture, therefore, may be found out from the known quantity of water vapor content in the atmosphere. Since water vapor is a polar molecule, the dipole moment is induced when a microwave propagates. Hence the water vapor molecules reorient themselves according to the polarity of propagation and cause a change in the refractive index of the atmosphere. This refraction is introduced in the time interval due to excess path length or delay (cm) or range error, ΔR. Since the refractive index of the troposphere at the surface of the earth is only 0.0003% greater than unity, the index of refraction is more conveniently expressed in terms of refractivity, N, as

$$N = (n-1) \times 10^6 \tag{1.14}$$

It is the excess over unity of refractive index expressed in millionths. Thus, at the surface, where $n = 1.000325$, the value of N is 325 units.

The range error is defined as

$$\Delta R = 10^{-6} \int_0^h N \, dh \tag{1.15}$$

The propagation delay has two main constituents. These are the dry path delay and wet path delay. The dry delay mainly depends on the amount of air through which the signal propagates. Hence it can be easily modeled with surface pressure measurements. The wet path delay depends on the precipitable water vapor through which the signal propagates.

However, the term *delay* refers to change in path length due to change in refractive index during the propagation of radio signals through the atmosphere duly constituted by several gases. Now due to the decrease in signal velocity there occurs an increase of time taken by the signal to reach to the receiving antenna (Adegoke and Onasanya 2008). Besides this, bending of the ray path also increases the delay (Collins and Langley 1998).

The refractivity as discussed is divided into two parts. The term N_h is the refractivity due to gases of air except water vapor and is called the hydrostatic refractivity. The term N_w is the refractivity due to water vapor and is called the wet refractivity. Hence from Equation (1.15) we rewrite

$$\Delta R = 10^{-6} \int N_h dh + 10^{-6} \int N_w dh \tag{1.16}$$

where $N_h = k_1 [\frac{P_d}{T}]$ and $N_w = k_2 [\frac{P_w}{T}] + k_3 [\frac{P_w}{T^2}]$.

The best average rather than best available coefficients provides a certain robustness against unmodeled systematic errors and increase the reliability of k values, particularly if data from different laboratories can be averaged. However, the available coefficients according to Ruger (2002) are given by

$$k_1 = 77.674 \pm 0.013 \, k/hpa$$

$$k_2 = 71.94 \pm 10.5 \, k/hpa$$

$$k_3 = 375406 \pm 3000 \, k^2/hPa$$

where P_d is the partial pressure due to gases, P_w is the water vapor pressure, and T is the ambient atmospheric temperature in kelvin. Hence the total refractivity N_r (ppm) is given by Maiti et al. (2009)

$$N_r = \frac{77.67P_d}{T} + 71.9P_w + 375406P_w / T^2 \qquad (1.17)$$

The contribution to dry component of atmospheric refractivity comes from an altitude from surface to lower stratosphere. The prediction of its variability although spatial and temporal is found to be easier in comparison to wet components of atmospheric refractivity, which shows the significant temporal and spatial variability depending on variation of atmospheric water vapor.

The variable parameters, namely temperature, pressure, humidity generally decreases with height and hence the radio refractivity gradient dN/dh usually is negative under standard atmospheric conditions. The effect of change in temperature on N values can further be explained by Figure 1.11. This suggests that at low temperature ranges, the change in humidity contributes a little to the change in N value. But, on the other hand, at temperatures higher than 20°C, the N value changes drastically with humidity. Therefore, it clearly indicates that changes in water vapor pressure are playing the major role in affecting the value of N. For all practical purposes, we take the liberty to consider the standard troposphere, which is extended up to 11 km, according to the International Commission for Navigation. The standard troposphere is considered to have a sea-level atmospheric pressure of about 1013 mb, sea level temperature of about 15°C, and relative humidity of about 60% with standard pressure lapse rate

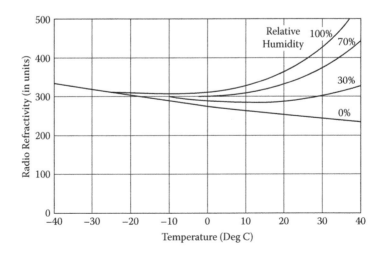

FIGURE 1.11
The radio refractivity of air as a function of temperature and relative humidity.

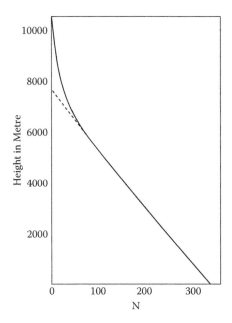

FIGURE 1.12
An idealized refractive index profile.

12 mb/100 m and standard temperature lapse rate 0.55 degree centigrade per 100 mt. Relative humidity remains unchanged at any height.

Now, the value of vertical refractivity gradient for standard troposphere is given by

$$\frac{dN}{dh} = -4.3 \times 10^{-2} m^{-1}$$

And the value of $\frac{dN}{dh}$ for well mixed air and with air temperature varying adiabatically is given by $\frac{dN}{dh} = -4.45 \times 10^{-2} m^{-1}$ which is in very close proximity to that obtained for standard troposphere.

Now, from the vertical profile of N, it follows that (Figure 1.12) the refractivity is constant throughout the atmosphere and it reduces to zero at about 8 km. The lapse rate of N begins to decrease from about 7.0 kilometers and then a bend appears in the profile. But in actual atmospheric conditions, temperature depends on weather conditions. There may be the more rapid changes due to the presence of local irregularities in the troposphere.

1.6.1 Forms of Refraction

The refractivity conditions are well defined in terms of the vertical refractivity gradient discussed in Section 1.6. A continuous state of turbulence in the

Forms of Refraction	Δ N Units/Km	'R'	a_e, Km	Actual Path	Equivalent Path
Sub Refraction	≥0	≤1	<6370		
Normal Refraction	0>To>−100	1<To<2.75	6370 To 17550		
Super Refraction	−100 To>−157	≥2.75	≥17550		
Ducting	≤−157	<0	<0		

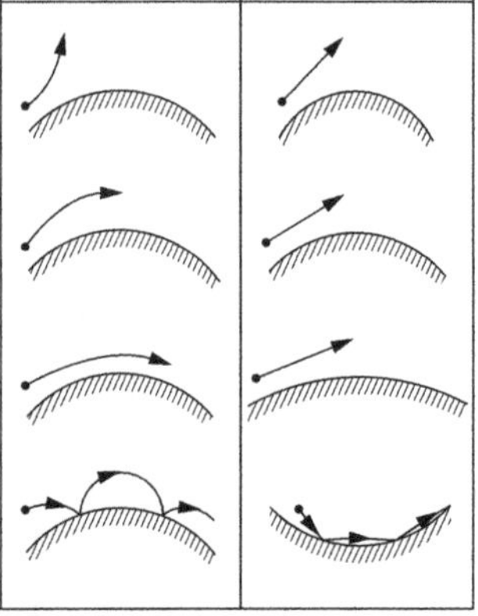

FIGURE 1.13
Various forms of atmospheric refraction.

atmosphere results in temporal and spatial variation in the values of *dN/dh*. The classification is done depending on the values of *dN/dh*, which is presented in Figure 1.13. This is essentially based upon the nature of bending of radio waves travelling through the troposphere with varying refractive index.

1.6.2 Normal Refraction (0 > Δ N > −100)

Usually refraction occurs in the atmosphere due to change in atmospheric density with height. Normally, refractive index decreases with height. Due to this change in refractive index, the waves are bent down instead of traveling in straight lines. This results in an increase of the radio-horizon path. What happens is that the top of the wave front travels in a rarer atmosphere than the bottom of the same wave front. Hence, in the rarer portion, the part of the wave front travels faster which results in a downward bent.

For standard atmosphere, the value of the effective earth radius factor is taken to be equal to 4/3 where

$$\frac{dN}{dh} \approx -40m^{-1} \text{ (approximately)}$$

1.6.3 Subrefraction (ΔN > 0)

When the value of $\frac{dN}{dh}$ is positive, that is, $\frac{dN}{dh}$ increases with height, the ray bends toward the normal. In Figure 1.13 it is shown that for the undeviated ray path the effective radius of the earth is less than the actual radius. In this case, line of sight propagation is decreased. Hence the value of R is less than one.

If the value of N equals to zero, then there is no change in refractivity, that is, $k = 1$ and hence there is no bending of rays. The rays travel in straight line. This is called a zero refraction condition.

1.6.4 Superrefraction ($-100 \geq \Delta N > -157$)

In the case of superrefraction, the bending of rays is away from the normal and bending is sharp (Figure 1.13). Hence, the LOS path is considerably larger.

1.6.5 Ducting ($\Delta N \leq -157$)

In the case of ducting, the ray path is more curved than the earth's surface. The rays emitted at a small angle of elevation undergo total internal reflection in the troposphere and return to the earth at some distances from the transmitter. Upon reaching the earth's surface and being reflected from it, the waves can skip further large distances. This situation results into a waveguide mode of propagation and is termed *ducting* (Figure 1.13). The conditions for ducting are taken for granted as an (a) abnormally high negative values of dN/dh, (b) extremely high lapse rate of humidity, and (c) occurrence of temperature inversion. Of these three factors, the most decisive one is the temperature inversion, which is due to change in dN/dh.

The inversion is again of two types: surface inversion (Figure 1.14) and elevated inversion (Figure 1.15). The elevated inversion occurs due to the advection process. This happens when a mass of warm air blows from the warmer region of earth over a cooled layer of air. This type of inversion normally occurs near the seacoast because warm dry air blows from land over the cooler sea. Another type is cooling of the earth's surface due to radiation affecting the lapse rate of humidity which may be extremely high. However, ducts can. form at ground level or at an elevated height depending on the terminal height at which the signal may or may not couple into the duct. To couple into and remain in a duct, the angle of incidence must be less than 1 degree. The depth of the duct and roughness are also important. If the duct depth is small compared to the wavelength, energy will not be trapped. If the roughness is small compared to the wavelength, energy will be scattered out of the duct. Surface ducts have the ground as a boundary and energy will be lost to the terrain, vegetation, and so forth. The significance of elevated ducts is that they can allow the signals to propagate for a long distance over the horizon.

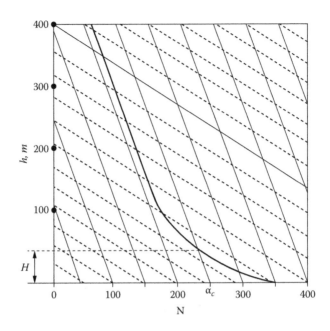

FIGURE 1.14
N = f(h) profile. The solids lines represent standard refraction and the broken lines represent critical refraction.

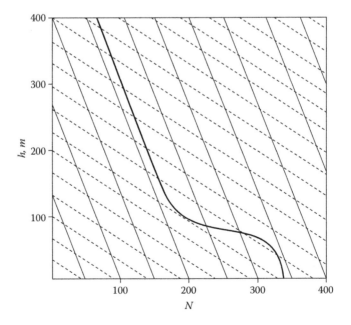

FIGURE 1.15
Refractive index profile in conjunction with elevated temperature inversion.

TABLE 1.3

Interaction between Atmospheric Constituents and
Electromagnetic Radiation

Agent	Effect	Magnitude	
		Shortwave	Longwave
CO_2, H_2O	Absorption	Moderate	Large
	Scattering	Small	Negligible
Aerosols	Absorption	Small	Small
	Scattering	Moderate	Small
Clouds	Absorption	Small	Large
	Scattering	Large	Large

1.7 Tropospheric Aerosols

Table 1.3 shows the effects of tropospheric aerosols on the radiation flux in
the atmosphere compared with those of gases and clouds. The particle effect
is found to be generally small except for aerosol. However, aerosol sizes of
0.02 to 0.2 m, which are likely to be attributable to human activities rather
than any natural process, may lead to variation in both the radiative changes
and cloud amount.

1.8 Rain Characteristics

At frequencies above 10 GHz electromagnetic radiation starts interacting
with the neutral atmosphere and with various meteorological parameters, in
particular the precipitation producing the absorption of energy of the incoming signal. This absorption of energy, in turn, depends on physical properties of rain. But because of complexity and the random behavior of physical
properties of rain, it is quite difficult to propose a universal model, to be
applicable everywhere in the world that is simple and sufficiently precise.

Rainfall of the tropics, where about 60% of global rainfall is concentrated,
is considered to be one of the major causes of climate change, global heat balance, and water cycle. In comparison to temperate climate, the tropical rain is
characterized by large frequency of occurrence and of bigger size. Karmakar
et al. (1991), in this regard, used the rapid response rain gauge, which has a
sensor consisting of mechanism in which rainwater is converted into rain
drops of constant volume and then the drops are collected electronically. It
has a collecting area of A ≈ 11 sq cm and capable of measuring rain intensity
up to 300 mm/hr with a time resolution of 10 sec.

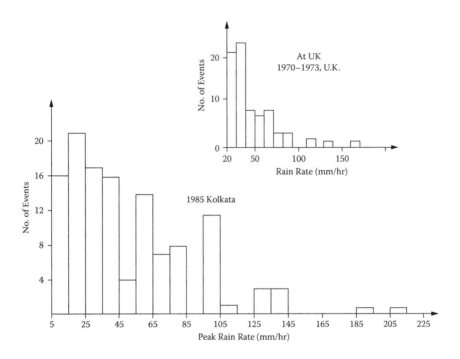

FIGURE 1.16
Histogram of duration of rain events over Kolkata and those over United Kingdom.

According to Lin (1975), the rain gauge integration time (T) to measure rain rate in a 1 m³ sampling volume (ΔV) is

$$T \approx \frac{\Delta V}{A v_r} \tag{1.18}$$

where v_r is the rain drop terminal velocity and is considered to be $v_r = 7$ mt/ sec. The dependence of rain rate distribution has been discussed in detail by Bodtman and Ruthroff (1974).

A statistics of peak rain rate at Kolkata (22° N) during 1985–1986 is shown in Figure 1.15. This is comparable to a similar histogram for a longer period of four years obtained for the measured peak rain rate in the United Kingdom (Norbury and White 1975) as shown in Figure 1.16. The distribution of duration of rain event at the same places is shown in Figure 1.17. These exhibit a similar trend in the values and of the form of the histogram, except few cases of very high rain rates occurring at Kolkata. The total number of events with rain rate exceeding 25 mm/hr is 85 measured at Kolkata, whereas in the United Kingdom the total number of events is only 77 for rain rates exceeding 20 mm/hr during the years 1970 to 1973. This exhibits only the large abundances of rain events in a tropical station. Note that the rain rate beyond 75 mm/hr over Kolkata is scanty. The cumulative rain rate distribution up to

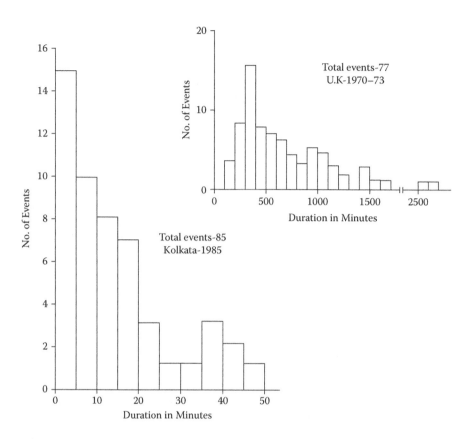

FIGURE 1.17

Statistics of peak rain rate over Kolkata (1985) and those over United Kingdom (1970–1973). Similarity of these two histograms is exhibited.

75 mm/hr is well fitted with normal distribution. But, on the other hand, taking the whole range of data set for Kolkata it was found that the lognormal distribution is well fitted for the years 1985 to 1987 and for 1991. To ascertain the situation of the rain rate up to 75 mm/hr the following examples may be cited. To get the proportion of rain rate beyond 50 mm/hr, let us take the mean value for the year 1985 and the deviation from the mean to the limit of the respective classes is $50 - 32.95 = 17.05$. Now, referring to the statistical table, the expected area is 0.3810 between mean value and 50 mm/hr. Hence the probability of occurrence of rain rate beyond 50 mm/hr is $0.5 - 0.38 = 0.12$ or 12%.

1.8.1 Raindrop Size Distribution Model

While considering the propagation of radio waves of frequency above 10 GHz through the rain environment, it becomes necessary to model raindrop size distribution. Several models have been proposed for the purpose but no definite conclusions of the model have been obtained regarding its suitability all over the country, which includes temperate and tropical climates. Most of the models were based on drop-size data from temperate regions of the world with exceptions of Ajayi and Olsen (1983, 1985). This is likely due to unavailable rain data from the tropical regions.

The basic distributions relate raindrop size distribution with certain limitations. The two types of distributions are (1) Laws and Parsons distribution (LP), and (2) Marshall and Palmer distribution (MP).

1.8.1.1 LP Distribution

The LP distribution has been favorably compared to measurements of stratiform and convective rain, and is probably the best known drop-size distribution. The International Radio Consultative Committee, while considering the propagation below 30 GHz, has recommended the LP distribution. But small raindrop diameters less than 1.0 mm were not properly described in this model. Hence its suitability is questionable for frequencies above 30 GHz. Since its publication, the distribution has appeared in a number of forms all of which describe the size of raindrops as a percentage of total volume (Setzer 1970).

The distribution $m(D)$ is defined as the fraction of total water volume collected for a given rainfall rate R per unit time, per unit area due to raindrops with diameter ranging between $D - \frac{\Delta D}{2}$ and $D + \frac{\Delta D}{2}$ where ΔD is 0.5 mm and D varies from 0.5 to 0.7 mm. The transformation from the distribution $m(D)$ into the distribution $N(D)$ as required by the theory is accomplished through the use of the relation

$$R_m(D) = 6\pi \times 10^5 D^3 V(D) N(D) \tag{1.19}$$

where, $V(D)$ is the drop terminal velocity and $N(D)$ is the number of drops per cubic meter (m^3).

1.8.1.2 Negative Exponential Distribution

The work of Marshall and Palmer (1948) resulted in an exponential distribution:

$$N(D) = N_o \exp(-\Lambda D) \tag{1.20}$$

where N is a constant and expressed in m^{-3} mm^{-1} and Λ is a coefficient in millimeters, which depends on rainfall rate.

In this regard, Fang and Chen (1982) present a list of values for N_0 and Λ in 33 different experiments by different authors for different places.

Marshall and Palmer proposed the values of $N_0 = 8000$ and $\Lambda = 4.1R$, where R is the rain rate in millimeters per hour (mm/hr). This distribution reproduces the average distribution obtained by Laws and Parsons quite well. But this distribution overestimates the number of smaller raindrops as it increases exponentially when D tends to zero. So the use of this distribution is suited only for frequencies above 30 GHz propagation. Below 30 GHz, this effect is not so important.

1.8.1.3 Gamma Distribution

This distribution is given by

$$N(D) = N_o D^{\alpha} \, \alpha \exp(-\Lambda\beta) \tag{1.21}$$

It is apparent from this distribution that it corrects the negative exponential distribution, which says the exponential increases with raindrop diameter per unit volume for D tending to zero. But the great difficulty lies with the use of this model. Ajayi and Olsen (1983) proposed a method for obtaining the parameters $N_o, \Lambda, \beta, \alpha$ but subsequently the same authors affirmed that they have found difficulty in accurately fitting the smaller raindrops to Gamma distribution (Ajayi and Olsen 1985). Instead, they fit the average drop-size spectra, measured in Nigeria, to a single lognormal distribution over the entire range of drop size, from drizzle to thunderstorm rain.

1.8.1.4 Lognormal Distribution

The tropical lognormal distribution is given by Zainel et al. (1993):

$$N(D) = \frac{N_T \exp\left[-0.5\left\{(\ln D_i - \mu)\big/\sigma\right\}^2\right]}{6D\sqrt{2\pi}} \tag{1.22}$$

where N_T is the total number of drops of all sizes, μ and σ are found to be the mean and standard deviation of D. Here μ and σ are found to be the functions of rain rate and depends on climate, geographical location, and rain type. They are presented as

$$N_T = a_0 R^{b_0}$$

$$\mu = A_\mu + B_\mu \ln(R) \tag{1.23}$$

$$\sigma^2 = A_\sigma + B_\sigma \ln(R)$$

Harden et al. (1978) presented the values of these parameters for Britain, and Fang and Chen (1982) presented values for Melbourne, Australia. Similarly, Verma and Jha (1996) presented values for Deharadun, India; Ajayi

TABLE 1.4

Values of Parameters in Equation (1.23)

Place	a_0	b_0	A	B	A	B
India	169.05	0.2937	−0.5556	0.13096	0.30042	−0.0236
Nigeria	108	0.363	−0.195	0.119	0.137	−0.013
Brazil	0.859	1.535	−0.023	0.116	0.805	−0.15

and Olsen (1985) for Nigeria; and Pontes et al. (1990) for Brazil. The parameters are presented in Table 1.4.

1.8.2 Raindrop Terminal Velocity

N (D) is the number of drops per unit volume having diameter D and D + dD at R mm/hr. According to Manabe et al. (1984), the rain rates (mm/hr) can be described as

$$R = 6\pi \times 10^5 \int_0^\infty V(D) N(D) D^3 dD \tag{1.24}$$

where $N(D)$ is the number of drops per unit volume having diameter D and $D + dD$ at R mm/hr and V (D) is the terminal velocity as was initially described by Gunn and Kinger (1949). However, later on at different times few analytic expressions were given for velocity of raindrops as

$$V(D) = 14.2 \, D0.5 \text{ (Spithans 1948)}$$

$$V(D) = 17.67 \, D^{0.67} \text{ (Atlas and Ulbrich 1977)}$$

$$V(D) = 9.25 - 9.25 \exp [-(6.8 \, D^2 + 4.88)] \text{ (Lhermitte 1988)}$$

$$V(D) = 9.65 - 10.3 \exp [-0.6D + 0.65 \, (-7D)] \text{ (Maitra and Gibbins 1994)}$$

According to the physical characteristics of convective and stratiform clouds, precipitation type can be identified with the help of simultaneous observations of vertical air velocities and terminal fall speeds of hydrometeors (Houze 1993). Until recently these observations had been rare (e.g., Yuter and Houze 1997; Atlas and Ulbrich 2000); indirect methods have been developed based on various observed properties of convective and stratiform clouds, mostly based on the reflectivity echo pattern as measured by surface weather radars (Churchill and Houze 1984; Steiner et al. 1995). However, due to interregional and interevent variation of the structure of rainfall-giving clouds, no general model for separation of type of rainfall has been found to be acceptable to all.

References

Adegoke, A. S., Onasanya, M. A., 2008, Effect of propagation delay on signal transmission, *Pacific Journal of Science and Technology (USA)*, 9, 1, 13–19.

Ajayi, G. O., Olsen, R. L., 1983, Measurements and analysis of rain drop size distributions in South Western Nigeria, *Proceedings of the URSI Commission F Symposium*, Louvain, Belgium ESA SP-194, 173–184.

Ajayi, G. O., Olsen, R. L., 1985, Modeling of a tropical rain drop size distribution for microwave and millimeter wave applications, *Radio Science*, 20, 193–202.

Atlas, D., Ulbrich, C. W., 2000, An observationally based conceptual model of warm oceanic convective rain in the tropics, *Journal of Applied Meteorology*, 39, 2165–2181.

Atlas, D., Ulbrich, C. W., 1977, Path and area integrated rainfall measurements by microwave attenuation in 1-3 cm band, *Journal of Applied Meteorology*, 16, 1322–1331.

Bodtman, W. F., Ruthroff, C., 1974, Rain attenuation on short radio paths: Theory and experiment, *Bell System Tech Journal*, 53, 1329–1349.

Buck, A. L., 1981, New equations for computing vapor pressure and enhancement factor, *Journal of Applied Meteorology*, 20, 1527–1532.

Churchill, D. D., Houze, R. A., 1984, Development and structure of winter monsoon cloud clusters on 10 December 1978, *Journal of Atmospheric Sciences*, 41, 933–960.

Clapeyron, E., 1834, Mémoires sur la puissance motrice de la chaleur, *Journal de l'École Polytechnique*, 14, 153–190. Also in *Annalen Physik Chem (Poggendorff)*, 59, 446–566, 1848.

Clausius, R., 1850, Ueber die bewegende Kraft der Wärme, *Annalen der Physik Chem* (Poggendorff), 79, 368–550.

Collins, P., Langlay, R. B. 1998, *Tropospheric propagation delay. How bad can it be?* Presented at the ION GPS-98. 11th International Tech Meeting of Satellite Division of ION. Nashville, Tennessee, Sept. 15–18.

Dolukhanov, M., 1971, *Propagation of radiowaves*. Moscow: Mir Publishers.

Fang, D. J., Chen, C. H., 1982, Propagation of centimeter and millimeter waves along a slant path through precipitation, *Radio Science*, 17(5), 989–1005.

Gibbins, C. J., 1990, A survey and comparison of relationships for the determination of the saturation vapor pressure over plane surfaces of pure water and of pure ice, *Annales Geophysicae*, 8(12), 859–886.

Goff, J. A., Gratch, S., 1946, Low-pressure properties of water from 160 to 212 F, *Transactions of the American Society of Heating and Ventilating Engineers*, 52, 95–121.

Gunn, K. L. S., Kinger, G. D., 1949, The terminal velocity of fall for water droplets in stagnant air, *Journal of Meterology*, 6, 243–248.

Harden, B. N., Norbury, W., White, J. K., 1978, Use of a lognormal distribution of raindrop sizes in millimetric radio attenuation studies, *IEEE Conference*, publication no. 169, part 2, 87–91.

Henderson-Sellers, B., 1984, A new formula for latent heat of vaporization of water as a function of temperature, *Quarterly Journal of the Royal Meteorological Society*, 110, 1186–1190.

Houze, R. A., Jr., 1993, *Cloud Dynamics*, San Diego, CA: Academic Press.

Hyland, R. W., 1975, A correlation for the second interaction virial coefficients and enhancement factors for moist air, *Journal of Research NBS*, 79A, 551–560.

Karmakar, P. K., Bera, R., Tarafdar, G., Maitra, A., Sen, A. K., 1991, Millimeterwave attenuation in relation to rain rate distribution over a tropical station, *International Journal of Infrared and Millimeter Waves (USA)*, 12, 1993–2001.

Lhermitte, R., 1988, Cloud and precipitation sensing at 94 GHz, *IEEE Transactions on Geoscience and Remote Sensing*, 26, 207–216.

Lin, S. H., 1975, A method of calculating rain attenuation distribution on microwave paths, *Bell System Technical Journal*, 54, 6, 1051–1085.

Maiti, M., Dutta, A. K., Karmakar, P. K., 2009, Effect of climatological parameters on propagation delay through the atmosphere, *The Pacific Journal of Science and Technology (USA)*, 10, 2, 14–20.

Maitra, A., Gibbins, C. J., 1994, *Inference of rain drop size distribution from measurements of rain fall rate and attenuation at infrared wavelength*, NRPP Research Note no. 140, Rutherford Appleton Laboratory (U.K.).

Manabe, T., Ihara, T., Furuhama, Y., 1984, Inference of rain drop size distribution from attenuation and rain rate measurements, *IEEE Transactions on Antenna and Propagation*, AP-32, 5, 254–256.

Marshall, J. S., Palmer, W. McK., 1948, The distribution of rain drops with size, *Journal of Meteorology*, 5, 165–166.

Norbury, J. R., White, W. J. K., 1975, Rain date distribution over U.K. *Meteorological Magazine*, 104, 221.

Pontes, M. S., Mello Silva, L.A.R., Migliora, C. S. S., 1990, K_U- Band slant path radiometric measurements at three locations in Brazil, *International Journal of Satellite Communication*, 8, 239–249.

Queney, P., 1974, *Elements de Météorologie*, Paris, Masson et Cie.

Ruger, M. J., 2002, *Refractive index formulae for radio waves*, 12th International Congress, Washington, DC, April 19–26.

Schelling, J., Burrows, C., Ferrel, E., 1933, Propagation of radio waves, *Proc IRE*, 21, 426–463.

Sen, A. K., Karmakar, P. K., 1988, Microwave communication parameters estimated from radiosonde observation over the Indian subcontinent, *Indian Journal of Radio and Space Physics*, 17, 165–171.

Setzer, D. E., 1970, Compound transmission at microwave and visible frequencies, *Bell System Tech Journal*, 49, 1873–1892.

Spithans, A. F., 1948, Drop size intensity and radar echo of rain, *Journal of Meteorology*, 5, 161–164.

Steiner, M., Houze, R. A. Jr., Yuter, S. E., 1995, Climatological characterization of three dimensional storm structure from operational radar and rain gauge data, *Journal of Applied Meteorology*, 34, 1978–2007.

Verma, A. K., Jha, K. K., 1996, Rain drop size distribution model for Indian climate, *Indian Journal of Radio and Space Physics*, 25, 15–21.

Yuter, S. E., Houze, R. A., 1997, Measurements of raindrop size distributions over the Pacific warm pool and implications for Z–R relations, *Journal of Applied Meteorology*, 36, 847–867.

Zaniel, A. R., Glover, I. A., Watson, P. A., 1993, Rain rate and drop size distribution measurements in Malasiya, *IGARSS-1993*, Aug, 309–311.

2

Propagation of Radio Waves: An Outline

2.1 Introduction

When electromagnetic wave propagates through the atmosphere and reaches the receiver, it takes its own path through a number of ways. Sometimes the wave arrives at the receiver after reflection from the ionosphere but ionospheric reflection is not being discussed in this book. There are certain others that may reach the receiver after reflection, the details of which will be discussed in Chapter 3. The waves may also reach the receiver after refraction and diffraction, which we will discuss in Chapter 4. The waves that arrive at the receiver after reflection from the troposphere and ionosphere are called sky waves. Waves being received by other paths, namely, space waves and surface waves, are called ground waves (Ghosh 1998). The space waves consist of waves reflected from the surface of the earth and also those received directly from the transmitter. The surface waves are guided along the earth's surface. These electromagnetic waves propagate close to the earth's surface and partly follow the curvature of the earth by diffraction.

Almost all the frequencies in the electromagnetic spectrum, for short-range communication, are propagated as ground waves. Waves shorter than 10 m may be treated as tropospheric waves. Ground waves are also treated as tropospheric waves. Propagation of ground waves is largely affected by electromagnetic properties of the earth's surface, namely, by its permittivity and conductivity. As the earth's surface is nonuniform and consisting of land, saline water of oceans and fresh water of lakes, the ground wave propagation is dependent on the electrical properties of the earth's surface. The ground wave is considerably affected by surface irregularities of the earth. For very low frequency (VLF) and low frequency (LF) waves with wavelengths in tens of kilometers, all types of irregularities except mountains are considered to be smooth. But in case of ultrahigh frequency (UHF) and super high frequency (SHF) between waves and low sea waves or even vegetation behave as a rough surface.

By the term *free space* we mean that the medium is homogeneous, nonscattering, and of unit dielectric constant. We consider the electromagnetic waves spread uniformly in all directions from the source since no obstacles are present in free space. The wave front is thus spherical in free space and will propagate outward in all directions with the speed of light. But to simplify the situation, we assume the transmitted waves are moving along a straight line, that is, "rays" are considered to propagate from the source in all directions and will continue to do so. It is to be remembered that energy carried by photons will eventually decay. But this decay of energy is not that serious. The rays are perpendicular to the tangential planes of the wave front. To visualize the concept of rays propagation, we consider an aerial that radiates equally in all directions. Such an aerial is called anisotropic radiator. This type of radiator is nonexistent, but to a close proximity, isotropic radiator may be replaced by a directional aerial that radiates energy in the desired direction. This type of aerial is best described by the term *power gain* (Collin 1982).

2.2 Power Gain of Directional Aerial

To get a clear idea about power gain we put two aerials, of which one is isotropic and the other one is directional, side by side. If the same power can be radiated by the aerials then the directional aerial will produce a stronger field at a particular distance than that of the isotropic radiator. This is because the isotropic radiator will radiate the power uniformly in all directions. The power fed to the isotropic radiator is increased to get the same field at the receiving end equal to that produced by the directional aerial. The factor A_p, showing how much power is fed to the isotropic radiator to get the same field as that produced by the directional aerial, is the power gain. So it is the ratio. That is to say, the power that must be radiated by an isotropic radiator to get a certain field strength at a certain distance divided by a particular power needed. It produces a ratio that we call power gain. This gain must not be confused with the directive gain. In case of power gain we have to consider the antenna efficiency, whereas in case of directive gain only the radiated power in a particular direction is considered. The power gain is expressed as

$$A_p = \eta D \tag{2.1}$$

where A_p is power gain; η is efficiency equal to 1, for lossless aerial; and D is directive gain (maximum). Sometimes, the efficiency of the antenna (directional) is found to be very close to 100%. For simplicity we shall consider that power gain and directive gain are the same.

2.3 Free Space Field Due to Directional Transmitting Aerial

Consider an isotropic radiator is placed in free space. This radiator is transmitting power P_t watt, which is received by a receiver at a distance r meters from the radiator. The power flux density at a distance r is given by

$$P_1 = \frac{P_t}{4\pi r^2} w/m^2 \tag{2.2}$$

The field strength produced by the isotropic radiator at a distance r is (see Equation 1.17)

$$E_{rms} = \frac{\sqrt{30 P_t}}{r} v/m$$

Thus, the field strength as produced by the directional aerial is found from the power gain G_t and the field produced by the isotropic radiator. It is written as

$$E_{rms} = \frac{\sqrt{30 P_t G_t}}{r} v/m \tag{2.3}$$

Thus the field strength is inversely proportional to the range and hence the power, which is proportional to the square of the field, is inversely proportional to the square of the range. Hence, the peak value of the field is

$$E_p = \frac{\sqrt{2}\sqrt{30 P_t G_t}}{r} = \frac{\sqrt{60 P_t G_t}}{r} v/m \tag{2.4}$$

Therefore, the instantaneous field is given by (consider only real part of $e^{i(wt-kr)}$)

$$E = E_p \cos(wt - kr) \tag{2.5}$$

where $k = \frac{2\pi}{\lambda}$, the wave number.
Therefore,

$$E = \frac{\sqrt{60 P_t G_t}}{r} \cos(wt - kr) \tag{2.6}$$

2.4 Power at the Receiving Directional Aerial

For all practical purposes of radio communication, we usually describe the propagation condition in terms of power produced at the input to the receiving system instead of field intensity. This needs the knowledge of power gain of the receiving aerial and the effective capture area of the aerial.

The power flux density at the receiver input is given by

$$P_2 = P_t G_t \tag{2.7}$$

where P_t is the power flux density at a distance r where the receiver is located (see Equation 2.2).

Therefore, the power received by the antenna

$$P_r = \frac{P_t G_t}{4\pi r^2} \tag{2.8}$$

where A is the effective capture area of the receiving antenna that is the area over which the antenna gathers the energy of the incoming radio wave. This is given by

$$A = \frac{G_r \lambda^2}{4\pi} \tag{2.9}$$

where G is the power gain of the receiving aerial and λ is the wavelength. From Equations (2.8) and (2.9), we write (Dolukhanov 1971)

$$P_r = \frac{P_t G_t G_r \lambda^2}{(4\pi r)^2} \tag{2.10}$$

This is the power received at the receiving terminal.

To illustrate this let us take an example of Voyager 1, which is at about 15 billion km from earth. It has a transmitter power of 13 watts at 8.415 GHz. The 3.7 m antenna has a gain of 48 dB. This produces an effective power toward the earth of about 800 kw. This signal is received by a large 70 m dish at Goldstone Mountain in Idaho. This dish has an area of 3800 m. So the total power it receives over that area is about 1×10^{-18} watts. It should be mentioned that Voyager 1 is now the most distant manmade object at 100 Au (1 Au is the radius of the earth) from the Sun and heading away at 3.6 Au per year. It is so far away that light takes nearly 14 hours to make the trip.

2.5 Free Space Transmission Loss

Free space loss consists of two effects. First is the effect of electromagnetic energy spreading in free space, which is determined by the inverse square law. This is given by Equation 2.2, which is rewritten as $P_1 = \frac{P_t}{4\pi r^2} w/m^2$. Note that this is not the frequency-dependent effect. The second effect is that of the receiving antenna's aperture, which describes how well an antenna can pick up power from the incoming electromagnetic wave. For an isotropic antenna this is given by

$$P_r = \frac{P_1 \lambda^2}{4\pi}$$

where P_r is the received power. Note that it is entirely dependent on wavelength. Now using the two aforementioned equations, we get propagation loss in free space (FSPL), which is mathematically written as

$$\text{FSPL} = \frac{P_t}{P_r}$$

Substituting the values of P_r from Equation (2.10), we write

$$L = P_t \frac{(4\pi r)^2}{P_t G_t G_r \lambda^2}$$

$$L = \frac{(4\pi r)^2}{\lambda^2} \times \frac{1}{G_t G_r}$$

(2.11)

If we use the isotropic radiator instead of directional antenna at the same place then, $G_t = G_r = 1$. And in that case Equation 2.11 takes the form

$$L_b = \frac{(4\pi r)^2}{\lambda^2}$$

This L_b is then defined as basic transmission loss. Here we see that L_b is independent of the aerials used. Now expressing Equation 2.11 in decibels and not in absolute units, we write

$$10 log_{10} L = 20 log_{10}(4\pi r) - 20 log\lambda - G_t(dB) - G_r(dB)$$

(2.12)

Here distance r and wavelength λ are expressed in meters.

2.6 Radio Waves in Neutral Atmosphere

When radio waves propagate through neutral atmosphere, power flux density decreases with distance covered. This is solely due to divergence of wave fronts. In other words, we can say that as the wave advances its energy is dispersed. For simplicity's sake, we consider that electromagnetic energy is received by the receiver at a large distance from the source. The transmitting antenna radiates energy in the form of waves and it expands as waves travel from the source and, at a large distance, the waves appear as uniform plane wave (Rao 1972).

Now we consider the wave is propagating in a medium that does not have free charges (neutral atmosphere) but has finite conductivity and hence the conduction current, $\bar{J}$. When radio waves propagate in such an imperfect conductor, currents are induced that cause absorption of energy. This absorption manifests as heat, and the radio waves get attenuated. The extent to which a radio wave is attenuated can be conveniently determined by using complex permittivity. From field theory it is known that when an electric field of a radio wave propagates in an ideal dielectric it is described as

$$E_z = E_m e^{i\omega\left(t-\frac{x\sqrt{\epsilon}}{c}\right)} \text{ V/m}$$

where ϵ is the ideal dielectric constant and $\frac{\sqrt{\epsilon}}{c}$ is the reciprocal of velocity of propagation. It is assumed that the wave is propagated in the direction of the X axis and the electric field is oriented in the direction of the Z axis in the right-handed system of coordinates shown in Figure 2.1.

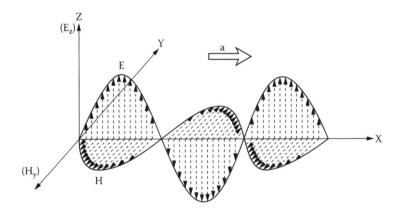

FIGURE 2.1
Electromagnetic field of a plane radio wave in an ideal dielectric.

Now, looking back to Maxwell's equation for a medium of finite conductivity and for time harmonic field, we write (Shevgaonkar 2006)

$$\nabla \times \vec{E} = -\mu \frac{\partial \vec{H}}{\partial t} = -j\omega\mu_0\mu_r\vec{H} \tag{2.13}$$

$$\nabla \times \vec{H} = \vec{J} + j\omega\epsilon\vec{E} = \vec{J} + j\omega\epsilon_0\epsilon_r\vec{E} \tag{2.14}$$

where μ_r and ϵ_r are relative permeability and relative permittivity of the medium, respectively. $\vec{J}$ is the conduction current. Again introducing Ohm's law, $\vec{J} = \sigma\vec{E}$, in Equation (2.14)

$$\nabla \times \vec{H} = \sigma\vec{E} + j\omega\epsilon_0\epsilon_r\vec{E}$$

$$= (\sigma + j\omega\epsilon_0\epsilon_r)\vec{E} \tag{2.15}$$

$$= j\omega\epsilon_0\left(\epsilon_r - \frac{j\sigma}{\omega\epsilon_0}\right)\vec{E}$$

Here, the term in the parenthesis is called the relative permittivity of the conducting medium and is written as (Sadiku 1995)

$$\epsilon_{rc} = \epsilon_r - \frac{j\sigma}{\omega\epsilon_0} \tag{2.16}$$

Hence, we see that the relative permittivity or alternatively dielectric constant is a frequency-dependent term. So, the propagation constant of the medium depends on frequency. The medium now may also behave as a conductor and a dielectric (see Equation 2.14). To a particular frequency, a medium may behave like a conductor and the same medium may be a dielectric to another frequency. This depends on relative contribution of the currents in two different situations. If the conduction current to a particular frequency dominates the displacement current then the medium will behave as a conductor to the given frequency. But on the other hand, if the displacement current is larger in comparison to the conduction current, then the medium will behave as a dielectric to the given frequency. This behavior is determined by the ratio of the conduction current density to the displacement current density, that is, $\sigma/\omega\sigma_0\sigma_r$. It is clear from this ratio that for a given electric field, the ratio is inversely proportional to frequency. Hence we are in a position to say that to the lower part of the electromagnetic spectrum, a given medium will behave like a conductor and for a higher frequency part the medium will behave like a dielectric.

For a uniform plane wave propagating in Z direction, the wave equation can be written as

$$\frac{\partial^2 E_x}{\partial Z^2} = -\omega^2 \mu_0 \epsilon_0 \epsilon_{rc} E_x = \gamma^2 E_x \tag{2.17}$$

where γ is the propagation constant and is given by

$$\gamma = \sqrt{-\omega^2 \mu_0 \epsilon_0 \epsilon_{rc}}$$
$$= j\omega\sqrt{\mu_0 \epsilon_0 \epsilon_{rc}} \tag{2.18}$$

$$= j\omega\sqrt{\mu_0 \epsilon_0} \sqrt{\epsilon_r - \frac{j\sigma}{\omega\epsilon_0}} \tag{2.19}$$

Therefore, the propagation constant is a complex quantity and can be written as

$$\gamma = \alpha + j\beta$$

$$\alpha = \omega\sqrt{\frac{\mu_0 \epsilon_0 \epsilon_r}{2}} \sqrt{\left\{\sqrt{1 + \sigma^2 / \omega^2 \epsilon_0^2 \epsilon_r^2} - 1\right\}} \tag{2.20}$$

and

$$\beta = \omega\sqrt{\frac{\mu_0 \epsilon_0 \epsilon_r}{2}} \sqrt{\left\{\sqrt{1 + \sigma^2 / \omega^2 \epsilon_0^2 \epsilon_r^2} + 1\right\}} \tag{2.21}$$

Here α is the real part of γ and termed as the attenuation constant, and β is the imaginary part, which is called the phase constant where β is related with the wavelength λ as

$$\lambda = \frac{2\pi}{\beta} \tag{2.22}$$

Therefore, we see from Equations 2.20 and 2.21 that α and β both increase with an increase in σ, that is, as the conductivity increases, the attenuation increases as well as the phase. So from Equation 2.22 it is clear that as β increases, the wavelength decreases. Hence, for a short wavelength, the phase constant is large with increasing conductivity. It is interesting to note that in a dielectric medium with finite conductivity, as the conductivity increases, the attenuation increases. So, a dielectric medium will produce attenuation if the conductivity is not zero and conversely a conductor has attenuation if the conductivity is not finite (Shevgaonkar 2006).

2.7 When Is a Medium a Conductor or Dielectric?

We know that in any medium having μ, σ, and ε properties, the magnitude of the ratio of conduction current density to the displacement current density (a nondimensional quantity) would be the determining factor for consideration of the electrical property of the medium. The conduction current is a characteristic of the conductor, whereas the displacement current is the characteristic of the dielectric. For a given electric field, if the conduction current is larger than the displacement current, the medium is considered to be a conductor and vice versa. The ratio $\sigma/\omega\epsilon$ is, however, a frequency dependent term and hence a conductor at one particular frequency may be a dielectric at the other frequency. For example, copper has $\sigma = 5.8 \times 10^7$ siemens/mt and $\epsilon = 9 \times 10^{12}$ farad/mt. Hence

$$\frac{\epsilon}{\omega t} \simeq \frac{10^{18}}{10^{16}} \text{ [setting } \omega = 10^{16} \text{ Hz]} \simeq 100$$

So up to 10^{16} Hz, copper behaves as a conductor but at a frequency of about 10^{20} Hz the value of the ratio, that is, $\sigma/\omega\epsilon$ becomes less or equal to 100 and hence copper behaves as a dielectric. But it should be emphasized that the ratio is inversely proportional to frequency. Hence, toward the lower part of the spectrum a medium mostly behaves as a conductor and toward the upper part of the spectrum the medium behaves like a dielectric.

Water is an example of varied frequency response. At low frequency, water is transparent to radiation and is of low loss. As frequency increases, the loss increases behaving as a dielectric loss. In the microwave region water behaves as a dielectric and happens to be lossy. It is extremely lossy to infrared making it opaque to radiation. The loss drops dramatically in the near infrared leading to the well-known transparent nature of water at visible frequency. Any dissolved chemicals such as salt alter the properties of water making it a lossy material to radio frequencies. For clarity, the electrical properties of importance in our study are given in Table 2.1.

TABLE 2.1

Electrical Properties of Different Types of Surfaces

Material	Value of ε (F/m)	Value of σ (S/m)
Sea water	80	04
Fresh water	80	10^{-3}
Moist soil	10	10^{-2}
Dry soil	04	10^{-3}
Forests	—	10^{-3}
Building material	—	7.5×10^{-4}
Mountains	—	7.5×10^{-4}

2.8 Wave Polarization

It is a common belief that electromagnetic wave is transverse in nature in which the electric vector and the magnetic vectors are perpendicular to each other and lie in a plane perpendicular to the direction of wave propagation. If the electric and magnetic vectors rotate in the transverse plane by same angle then the transverse nature of the electromagnetic wave will not be affected under any circumstances. However, the magnetic vector is specified by the knowledge of the electric field vector. Therefore, it is customary to discuss the rotation of the electric field vector. Thus the rotation of the tip of the electric field vector is discussed as wave polarization. In a broader sense, the polarization of an electromagnetic wave describes the orientation of the electric field vector at a given point in space during one complete oscillation. Depending on the orientation of the electric field vector, the polarization of the wave is described. In other words that the temporal behavior of the electric field vector gives the idea of polarization.

To illustrate the different states of polarization, we consider an electric field vector **E** propagating along the Z direction. This instantaneous electric field vector may have two components, E_1 and E_2, along the x and y direction, respectively. Let the two components be

$$E_1 = E_x \cos \omega t \tag{2.23}$$

$$E_2 = E_y \cos(\omega t + \varphi) \tag{2.24}$$

where ω and φ are the frequency of the wave and phase difference, respectively. From Equation (2.23), we write

$$\cos \omega t = \frac{E_1}{E_x} \tag{2.25}$$

$$\sin \omega t = (1 - \cos^2 \omega t)^{1/2} = \left[1 - \left(\frac{E_1}{E_x}\right)\right]^{1/2} \tag{2.26}$$

From Equation (2.24), we get

$$E_2 = E_y \cos \omega t \cos \varphi - E_y \sin \omega t \sin \varphi$$

$$= E_y \frac{E_1}{E_x} \cos \varphi - E_y \sqrt{1 - \frac{E_1^2}{E_x^2}} \sin \varphi$$

(refer to Equations 2.25 and 2.26)

or

$$E_y \sqrt{\left(1 - \frac{E_1^2}{E_x^2}\right)} \sin \varphi = E_y \frac{E_1}{E_x} \cos \varphi - E_2$$

or

$$\left(1 - \frac{E_1^2}{E_x^2}\right) \sin^2 \varphi = \left(\frac{E_1}{E_x} \cos \varphi - \frac{E_2}{E_y}\right)^2$$

or

$$\left(1 - \frac{E_1^2}{E_x^2}\right) \sin^2 \varphi = \frac{E_1^2}{E_x^2} \cos^2 \varphi + \frac{E_2^2}{E_y^2} - \frac{2E_1 E_2}{E_x E_y} \cos \varphi$$

or

$$\sin^2 \varphi - \frac{E_1^2}{E_x^2} \sin^2 \varphi = \frac{E_1^2}{E_x^2} \cos^2 \varphi + \frac{E_2^2}{E_y^2} - \frac{2E_1 E_2}{E_x E_y} \cos \varphi$$

or

$$\sin^2 \varphi = \frac{E_1^2}{E_x^2} - \frac{2E_1 E_2}{E_x E_y} \cos \varphi + \frac{E_2^2}{E_y^2} \tag{2.27}$$

Equation (2.27) describes the equation of ellipse and the wave is called elliptically polarized wave. The tip of the electric vector will draw an ellipse over one complete time period. But, it is to be remembered that depending upon the values of E_x, E_y, and φ, the shape and orientation of the ellipse will change.

2.8.1 Linear Polarization

If the phase difference between the two components is zero, that is, $\varphi = 0$, then from Equation (2.27), we have

$$\left(\frac{E_1^2}{E_x^2} - \frac{E_2^2}{E_y^2}\right)^2 = 0$$

or

$$\frac{E_1}{E_x} = \frac{E_2}{E_y}$$

or

$$E_2 = \frac{E_y}{E_x} E_1 \qquad\qquad (2.28)$$

Equation (2.28) describes the equation of a straight line with a slope E_2/E_y and the wave is called linearly polarized wave. Hence for $\varphi = 2\pi$, 4π, and so forth, the slope will remain E_2/E_y and Equation (2.28) will repeat itself. But for $\varphi = \pi$, 3π, and so forth, we obtain

$$E_2 = -\frac{E_y}{E_x} E_1 \qquad\qquad (2.29)$$

Equation (2.29) also represents a straight line passing through the origin but with opposite slope. If in Equations (2.28) and (2.29), we put $E_x = 0$, then the wave becomes vertical and the wave is called vertically polarized. If $E_y = 0$ then the wave is called horizontally polarized.

2.8.2 Circular Polarization

If the phase difference becomes $\varphi = \pm \pi/2$ and the amplitudes of the component waves become equal, that is, $E_x = E_y = E_z = 0$, then from Equation (2.27), we write,

$$\frac{E_1^2}{E_0^2} + \frac{E_2^2}{E_0^2} = 1$$

or

$$E_1^2 + E_2^2 = E_0^2 \qquad\qquad (2.30)$$

Equation 2.30 thus represents a circle and hence the wave is called circularly polarized. It is to be noted that the presence of a positive and negative sign in the value of φ should not alter the shape of the circle but it affects how the circle has been traced with time. Moreover, the sense of rotation has been determined by the sign of the phase term. If φ is positive then we normally say the rotation is clockwise, whereas for a negative sign we say the rotation is counterclockwise. In fact, this sense of rotation should depend on from where we see the rotation, that is, from which side (front or back) of the plane we are looking. If the wave motion is perpendicular to the plane of the paper, then the wave seems to be receding from us. So if the rotation is counterclockwise, then the sense of polarization is right-handed polarization. On the other hand, if the rotation is clockwise, then the sense of polarization is left-handed polarization.

References

Collin, R. E., 1982, *Antenna and radio wave propagation*, New York: McGraw-Hill.

Dolukhanov, M., 1971, *Propagation of radio waves*, Moscow: Mir Publisher.

Ghosh, S. N., 1998, *Electromagnetic theory and wave propagation*, New Delhi: Narosa.

Rao, N. N., 1972, *Basic electromagnetics with application*, Englewood Cliffs, NJ: Prentice Hall.

Sadiku, M. N. O., 1995, *Elements of electromagnetics*, Oxford: Oxford University Press.

Shevgaonkar, R. K., 2006, *Electromagnetic waves*, New Delhi: Tata McGraw-Hill.

3

Reflection and Interference of Radio Waves

3.1 Introduction

When discussing free space propagation we have not taken into account the presence of ground, mountains, buildings, seas, lakes, rivers, forests, and so forth. These should invariably be considered in the lowermost part of the atmosphere. Mostly 70% of the globe's surface is covered with water. The electrical property of water is very much different from that of land and because of this, the propagation of electromagnetic waves will be affected in a different way. Even water is not always the same in that respect; the saline water of seas and oceans are unlike the fresh water of lakes and rivers. Moreover, in the absence of wind, the surface of water will behave like a smooth surface whereas a sea surface is mostly rippled. This phenomenon will also affect the propagation. Land also has a wide variety of states, such as damp soil and dry soil. Plains are also covered with bushes, forests, barren rocks, and mountains.

Depending on surface irregularities, the propagation characteristics of radio waves vary with wavelength and hence with frequency. Thus the ground, mountains, buildings, and so on will reflect waves. Waves will be refracted as they pass through the atmosphere. Waves may also be diffracted around the tall and sharp rises of mountains, buildings, and so on. Radio waves with a frequency range of 3 to 300 KHz may run smoothly for tens of kilometers over all kinds of earth's surfaces except the effects for the presence of tall buildings and mountains. In contrast, microwaves (30–300 GHz) will have a rough going over sea surface, vegetation, and so forth. But in this chapter we will be discussing only the effect of reflection and interference in the propagation path. The other effects will be discussed in subsequent chapters.

3.2 Reflection of Radio Waves

Reflection of light by a mirrorlike smooth surface and the reflection of radio waves by a conducting medium exhibit a close resemblance. The well-known laws of reflection hold good in the case of reflection of radio waves. We know that the intensity of light waves gets reduced after reflection and the same thing happens in the case of radio waves. The intensity of radio waves gets reduced due to absorption at each reflection. With a perfect conductor, we should not expect any absorption of energy of radio waves. But neither earth's crust nor anything else is a perfect conductor. Therefore, essentially the imperfect conductor will give rise to absorption of energy although reflection is taking place from the surface. After the absorption, to some extent the remaining part will be transmitted and propagated through the medium accompanied by reflection whose effects will be discussed later. The important points regarding the reflection of radio waves to be noted are

1. Obstruction dimensions are very large compared to the wavelength of the radio signal used.
2. The intensity of the reflected wave is determined by the reflection coefficient, defined as the ratio of electric intensity of reflected wave to that of the incident wave, depending on frequency.
3. Electric vector is perpendicular to the conducting surface; otherwise surface current will set up and subsequently no reflection will take place.
4. Reflection of rays from different surfaces may interfere constructively or destructively at the receiver input, which we call multipath propagation.
5. Most surfaces are neither perfectly smooth nor rough.

According to Rayleigh, a surface is smooth if

$$\Delta h \sin\theta < \frac{\lambda}{8} \tag{3.1}$$

where Δh is the height difference between any two points on the surface, λ is the wavelength of the incident wave, and θ is the grazing angle.

The waves reflect from any two points on the surface at a height difference (Δh), which will be shifted in phase with respect to each other by an amount

$$\Delta\varphi = 4\pi\Delta h \frac{\sin\theta}{\lambda} \tag{3.2}$$

The radio wave will be reflected from a surface obeying the laws of reflection provided it obeys the Rayleigh roughness criterion (Equations 3.1 and 3.2). According to Rayleigh criterion, the surface might appear smooth for microwaves even though in sea the peak-to-trough wave height is large.

3.3 Plane Wave at Dielectric Interface

Let us assume the space is divided into two regions where the electric properties are characterized by μ and ϵ but with $\sigma = 0$ (Figure 3.1). In Figure 3.1, the dielectric interface is assumed to lie along the x–y plane passing through $z = 0$. We assume the wave vector $\vec{k}$ lies in the x–z plane and makes an angle θ_i with the normal to the interface, which lies in the z direction. Thus, θ_i is called the angle of incidence and the plane containing the wave vector $\vec{k}$ and the normal to the interface is called the plane of incidence. So, the angles that the wave vector makes with three axes are

$$\theta_x = \frac{\pi}{2} - \theta_i$$

$$\theta_y = \frac{\pi}{2} \tag{3.3}$$

$$\theta_z = \theta_i$$

The field equation for the incoming wave is written as

$$\vec{F}_i = \vec{F}_{i\theta} e^{-j\vec{k}\cdot\vec{r}}$$

$$= \vec{F}_{i\theta} e^{-j\beta_1(x\cos\theta_x + y\cos\theta_y + z\cos\theta_z)} \tag{3.4}$$

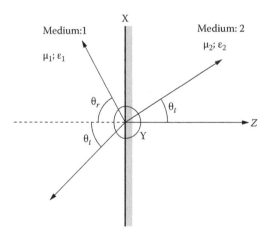

FIGURE 3.1
Two separate medium with different electrical properties.

where $\vec{F}_{i\theta}$ is the amplitude and β_1 is the phase constant of the wave in the medium 1. For all kinds of fields, a time variation of $e^{j\omega t}$ is implicitly assumed and hence not mentioned in our derivations presented subsequently.

From Equations (3.3) and (3.4), we get on substitution of $\theta's$

$$\vec{F}_i = \vec{F}_{i\theta} e^{-j\beta_1\left[x\cos\left(\frac{\pi}{2}-\theta_i\right)+y\cos\frac{\pi}{2}+z\cos\theta_i\right]}$$

$$= \vec{F}_{i\theta} e^{-j\beta_1(x\sin\theta_i+z\cos\theta_z)} \tag{3.5}$$

So it is clear from Equation (3.5) that the field vector is independent of the y direction, which means that at the interface, that is, at $z = 0$, the phase is constant along the y direction and increases linearly along the x direction. The magnitude of the field at the interface is now given (see Equation 3.5) by putting $z = 0$

$$|\vec{F}| = Real\{F_i e^{-j\beta_1 x\sin\theta_i}\}$$

$$= F_{i\theta} \cos(\beta_1 x\sin\theta_i) \tag{3.6}$$

Thus, the field at the interface is sinusoidal in nature. The same type of field would be induced on the other side of the interface due to the boundary conditions, which require maintaining continuity of the fields at the boundary. But the magnetic and electric field vectors in general cannot satisfy the boundary conditions simultaneously without modifying the incident field. So we say that when a plane wave is incident on the interface, the fields with similar phase variation are induced on both sides of the interface (Shevgaonkar 2006). Hence in region 1, the total field is the combination of the incident field and the induced field. But in region 2, the only field is the induced field.

The induced field in medium 1 is then bounced back into medium 1 from the interface, whereas the induced field in medium 2 will be going away from the interface. These waves will be as plane waves and, in medium 1 and medium 2, we have the corresponding reflected wave and transmitted (refracted) wave as shown in Figure 3.2. Again, referring to Figure 3.1, let θ_r be the angle of reflection and θ_t be the angle of refraction with respect to the interface normal. So, similar to Equation (3.3), we have the following sets of equations.

For the reflected wave

$$\theta_x = \frac{\pi}{2} - \theta_r$$

$$\theta_y = \frac{\pi}{2} \tag{3.7}$$

$$\theta_2 = \pi - \theta_r$$

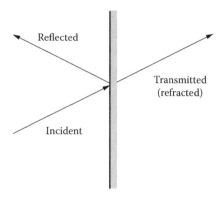

FIGURE 3.2
Reflected and transmitted components are shown separately.

And, for transmitted wave

$$\theta_x = \frac{\pi}{2} - \theta_t$$

$$\theta_y = \frac{\pi}{2} \tag{3.8}$$

$$\theta_z = \theta_t$$

The field equation for the reflected wave can then be written as (see Equation 3.4)

$$\begin{aligned}
\vec{F}_r &= \vec{F}_{r\theta} e^{-j\beta_1(x\cos\theta_x + y\cos\theta_y + z\cos\theta_z)} \\
&= \vec{F}_{r\theta} e^{-j\beta_1\left[x\cos\left(\frac{\pi}{2}-\theta_r\right) + y\cos\frac{\pi}{2} + z\cos(\pi-\theta_r)\right]} \\
&= \vec{F}_{r\theta} e^{-j\beta_1(x\sin\theta_r - z\cos\theta_r)}
\end{aligned} \tag{3.9}$$

Similarly, for the transmitted wave

$$\vec{F}_t = \vec{F}_{t\theta} e^{-j\beta_2(x\sin\theta_t + z\cos\theta_t)} \tag{3.10}$$

Equations (3.5), (3.9), and (3.10) represent the incoming, reflected, and refracted field equations, respectively. But at the interface, that is, at $z = 0$, the field equations must satisfy the boundary conditions. So we say that the tangential components of the electric field vector and normal components of the magnetic field vector must be continuous. Now, taking the electric field vector into account, we write that the sum of the tangential component of field vectors in medium 1 must be equal to the tangential component of the field vector in medium 2. This is possible only when the phases of the incident, reflected, and transmitted field vectors are the same.

So,

$$\beta_1 x \sin\theta_i = \beta_1 x \sin\theta_r = \beta_2 x \sin\theta_t \quad \text{for } z = 0;$$

or

$$\sin\theta_i = \sin\theta_r$$

or

$$\theta_i = \theta_r$$

Therefore, the angle of incidence is equal to the angle of reflection (law of reflection).

Again,

$$\beta_1 x \sin\theta_i = \beta_2 x \sin\theta_t$$

where, β_1 and β_2 are the phase constants for medium 1 and medium 2, respectively, and are given by

$$\beta_1 = \omega\sqrt{\mu_1\epsilon_1} \quad \text{and} \quad \beta_2 = \omega\sqrt{\mu_2\epsilon_2}$$

$$\therefore \omega\sqrt{\mu_1\epsilon_1} \sin\theta_1 = \omega\sqrt{\mu_2\epsilon_2} \sin\theta_2$$

For dielectrics, we write

$$n = \text{refraction index} = \sqrt{\mu\epsilon}$$

$$\therefore n_1 \sin\theta_1 = n_2 \sin\theta_2$$

So far, we have discussed the propagation of plane waves across the dielectric interface with any arbitrary orientation. But next we will discuss some specified orientations. The cases are

Horizontal polarization—When the **E** lies in the plane of incidence

Vertical polarization—When the **E** lies in a plane perpendicular to the plane of incidence

We generally talk about the case of vertical polarization. If the polarization is vertical we define two parameters and they are

Reflection coefficient $|R| = \frac{|E_r|}{|E_i|}$

Transmission coefficient $|T| = \frac{|E_t|}{|E_i|}$

It may also be proved that $|R|$ and $|T|$ are real which means that there should be no arbitrary change of phase after reflection or transmission of wave. Depending on the sign before R and T, the phase change could be either zero or π. Magnitude of reflection coefficient should always be less than one but the transmission coefficient may be greater or lesser than unity.

3.4 Reflection Coefficient for Flat Smooth Earth

The reflection from the smooth earth can be determined from Fresnel's equation, from which it is seen that the magnitude and phase of reflection depends on frequency, polarization, angle of incidence, and the electrical properties of the earth's surface.

Let us now assume that the electromagnetic wave is propagating through free space and is incident on a dielectric that is nonmagnetic. Then according to Figure 3.3, assuming that the electromagnetic wave

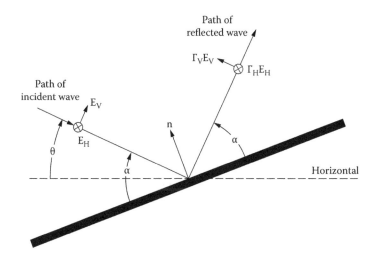

FIGURE 3.3
Reflection by a smooth plane (dielectric). $R_V E_V$ and $R_H E_H$ are the horizontal and vertical polarization, respectively.

on the dielectric is propagating in free space and the dielectric surface is nonmagnetic (Long 1983), the magnitude of reflection coefficient ρ, for horizontal and vertical polarization and from smooth surface, is given by, according to Fresnel

$$R_H = \frac{\sin\alpha - (\epsilon - \cos^2\alpha)^{1/2}}{\sin\alpha + (\epsilon - \cos^2\alpha)^{1/2}}$$

$$= \rho_H e^{-j\varphi_H} \text{ and}$$

$$R_V = \frac{\epsilon\sin\alpha - (\epsilon - \cos^2\alpha)^{1/2}}{\epsilon\sin\alpha + (\epsilon - \cos^2\alpha)^{1/2}}$$

$$= \rho_V e^{-j\theta_H}$$

where α is measured perpendicular to the surface normal *n*, and ε is the complex dielectric constant, which is, in turn, given by

$$\epsilon = \frac{K}{\epsilon_0} - \frac{j\sigma}{\omega\,\epsilon_0}$$

$$= \epsilon' - j\epsilon''$$

where ϵ_0 is the dielectric constant for free space, and $\omega = 2\pi f$, f is the propagation frequency. Here, the conductivity σ and permittivity K depend on frequency.

For normal incidence, $\alpha = \pi/2$, the two polarizations become equal and then

$$R_H = \frac{1 - \epsilon^{1/2}}{1 + \epsilon^{1/2}}$$

$$R_V = -\frac{1 - \epsilon^{1/2}}{1 + \epsilon^{1/2}} = -R_H$$

The negative sign indicates that the true field is reversed compared to the assumed field; R_H and R_V must therefore be different in sign if the reflected fields have the same phase. Figures 3.4 through 3.9 (Povejsil 1961) show the variation of ρ and φ versus wavelength λ and angle α for typical land and water surfaces. The curves show that for horizontal polarization the variation in magnitude of reflection coefficient and its corresponding phase changes are very little. This is caused by increased transmission into the surface that occurs near the Brewster angle, α_B. This happens when ρ_v reaches a minimum and $\varphi_v = \pi/2$ itself also depends on frequency. On the other hand, for

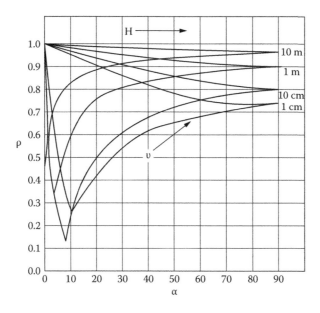

FIGURE 3.4
Magnitude of reflection coefficient for sea water at temperature 10°C as a function of incidence angle.

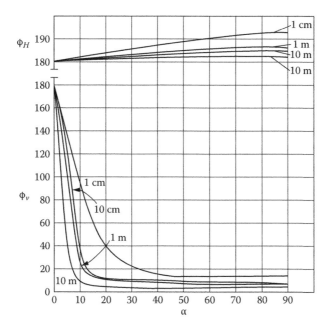

FIGURE 3.5
Phase of reflection coefficient for sea water at temperature 10°C as a function of incidence angle.

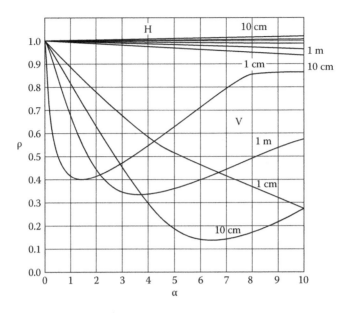

FIGURE 3.6
Same as Figure 3.4 but incidence angle is varied up to 10°.

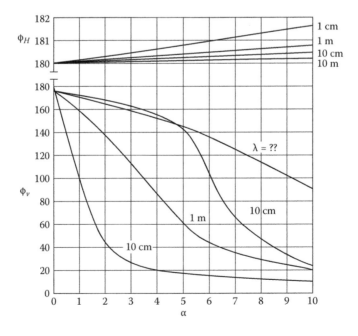

FIGURE 3.7
Same as Figure 3.5 but incidence angle is varied up to 10°.

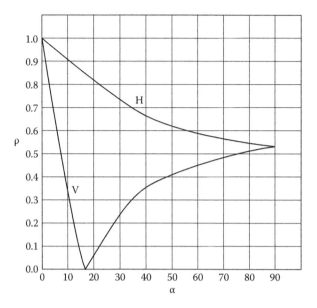

FIGURE 3.8
Magnitude of reflection coefficient for average land ($\epsilon = 10$, $\sigma = 1.6 \times 10^{-3}$ mho/m) as a function of incidence angle.

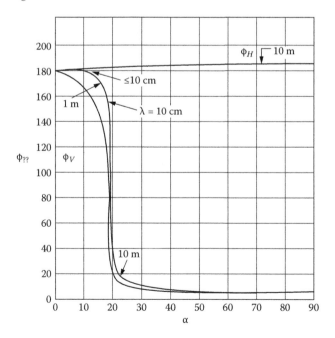

FIGURE 3.9
Phase of reflection coefficient for average land ($\epsilon = 10$, $\sigma = 1.6 \times 10^{-3}$ mho/m) as a function of incidence angle.

TABLE 3.1

Electromagnetic properties of soil and water

Medium	λ	σ (mho/m)	ϵ'	ϵ''
Sea Water	3m-20cm	4.3	80	774
20^0 -25^0C	10 cm	6.5	69	39
28^0C	3.2 cm	16	65	30.7
Distilled water, 28^0C	3.2 cm	12	67	23
Fresh water	1m	10^{-3}-10^{-2}	80	0.06
Dry Sandy Loam	9cm	0.03	2	1.62
Wet Sandy Loam	9cm	0.6	24	32.4
Dry Ground	1m	10^{-4}	4	0/006
Moist Ground	1m	10^{-2}	30	0.6

vertical polarization these variations are noticeable. It is to be noted that although the dielectric properties of water and land depend on incidence angle and wavelength, the values of φ_H and ρ_H can be approximated by π and unity, respectively. For small angle of incidence, φ_V and ρ_V can also be approximated by π and unity, but for large angles of incidence φ_V can be approximated to zero. The angles φ_H and φ_V differ negligibly from 180° for microwaves and depression angle of one degree or less. However, for clarity, the approximate electromagnetic properties of soil and water are given in Table 3.1 (Kerr 1951).

3.5 Field Strength Due to Reflection from Flat Earth

If the transmitting and the receiving aerials are a short distance apart, it is customary to consider that the earth is flat instead of spherical. Hence we assume that the radio waves are propagated along a flat, imperfectly conducting surface. It may further be assumed that the surface is smooth and uniform over the entire length of the propagation path. With the elevated transmitting and receiving aerials, the rigorous treatment of using Maxwell's equations to get the reflected field may be supplemented by using a simpler approach through geometrical approach. But before going any further, let us define the term *elevated aerial*. This implies that (a) the aerial uses a nonradiating feeder, that is, a very tall aerial with a radiating down lead will not be classified as elevated; and (b) the aerial height must be several times the wavelength (Dolukhanov 1971).

According to Figure 3.10, we write

r = Distance between the aerials

h_1 = Height of the transmitting aerial

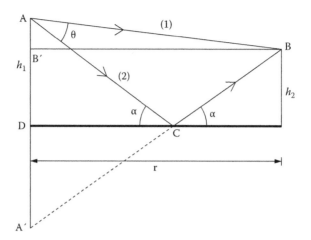

FIGURE 3.10
Radio wave path with horizontal polarization.

h_2 = Height of the receiving aerial
G_1 = Gain of the transmitting aerial
G_2 = Gain of the receiving aerial
P_t = Transmitted power
ϵ = Dielectric constant of the reflecting surface
σ = Conductivity of ground
λ = Wavelength used

Let us now rewrite Equation 2.3 as the received field in a more convenient way by expressing the power in kilowatts, the distance in kilometers, and the field in millivolts per meter, as

$$E_{rms} = \frac{173\sqrt{P_t G_t}}{r} F \ \text{mV/m} \ (F = \text{attenuation function}) \tag{3.11}$$

In other words, the received power is

$$P_r = \frac{6.33 \times 10^3 \, P_t G_1 G_2 \lambda^2}{r^2} \ nW$$

and the peak value of the field is

$$E_p = \frac{245\sqrt{P_t G_t}}{r} \tag{3.12}$$

According to Vedensky (Dolukhanov 1971), for the first time, the electric field of a radio wave at the reception point may be treated as a sum of the direct ray and ground reflected ray picked up by the receiving aerial. It is clear from Figure 3.10 that only one reflected ray is being received by the aerial. Thus the resultant field may be the sum of two instantaneous fields among which one is for the direct ray and the other one is for the reflected ray.

The instantaneous field as received by the receiving aerial is

$$\vec{E}_1 = \frac{245\sqrt{P_t G_t}}{r_1} e^{j\omega t} \tag{3.13}$$

And the other due to ground reflection is

$$\vec{E}_2 = \frac{245\sqrt{P_t G_t}}{r_2} e^{j\omega t - kr} \tag{3.14}$$

Replacing k with $2\pi/\lambda$ and r with $r + \Delta r$, we write

$$\vec{E}_2 = \frac{245\sqrt{P_t G_t}}{r + \Delta r} e^{j(\omega t - \frac{2\pi}{\lambda}\Delta r)} \tag{3.15}$$

But here it is to be remembered that we have restricted ourselves to the case where $h_1 \ll r$ and $h_2 \ll r$, that is, we have considered only the ground path. Another point to be remembered is that the direct ray and the reflected ray follows the same direction and that is why the gain terms in Equations (3.13) and (3.15) are the same for both the aerials (Figure 3.10).

Here, $r_1 = AB$ and $r_2 = AC + CB$ and $\Delta r = AC + CB - AB =$ difference in path length between the reflected and direct ray; R is the reflection coefficient (Figure 3.10). We further assume that h_1 and h_2 are much less than r and are a small fraction of horizontal distance r. It may be pointed out also from Equations (3.13) and (3.15) that since the path length of the reflected ray is longer by Δr, there appears a phase shift of $(\frac{2\pi}{\lambda})\Delta r$. Now, putting the reflection coefficient $\vec{R} = \bar{R}e^{j\theta}$ in Equation (3.15), we get

$$\vec{E}_2 = \frac{R245\sqrt{P_t G_t}}{r + \Delta r} e^{j(\omega t - \theta - \frac{2\pi}{\lambda}\Delta r)} \tag{3.16}$$

Now comparing Equations (3.13) and (3.16), we see that the amplitude and phase of reflected ray has been changed. The change in amplitude is caused due to loss in energy after reflection and change in phase is due to (a) shifting of phase by θ in reflection and (b) retardation in phase due to difference in path length. The resultant field at the point of reception may now be obtained by summing up the fields due to direct ray and ground reflected ray. Therefore, adding Equations (3.13) and (3.16), we get the resultant field at the point of reception

$$\vec{E} = \vec{E}_1 + \vec{E}_2$$

$$= \frac{245\sqrt{P_tG_t}}{r}\{1 + Re^{-j\beta}\}e^{j\omega t}$$

(3.17)

where $\beta = \theta + (\frac{2\pi}{\lambda})\Delta r$ (neglecting Δr in comparison to r in denominator of Equation 3.16).
Again,

$$1 + Re^{-j\beta} = 1 + R\cos\beta - jR\sin\beta$$

$$= \sqrt{1 + 2R\cos\beta + R^2}\,e^{-j\varphi}\,.$$

where

$$\tan\varphi = \frac{R\sin\beta}{1 + R\cos\beta}$$

Therefore, putting these values in Equation (3.17) and recalling that $\beta = \theta + \frac{2\pi}{\lambda}\Delta r$, we write

$$\vec{E} = \frac{245\sqrt{P_tG_t}}{r}\sqrt{1 + 2R\cos\left(\theta + \frac{2\pi\Delta r}{\lambda}\right) + R^2}\,e^{j(\omega t - \varphi)} \text{ mV/m}$$

(3.18)

and the root mean square (rms) value of E is

$$\vec{E}(rms) = 173\frac{\sqrt{P_tG_t}}{r}\sqrt{1 + 2R\cos\left(\theta + \frac{2\pi\Delta r}{\lambda}\right) + R^2} \text{ mV/m}$$

(3.19)

Now comparing Equations (3.19) and (3.11), we see that the attenuation function takes the form

$$F = \sqrt{1 + 2R\cos\left(\theta + \frac{2\pi}{\lambda}\Delta r\right) + R^2}$$

(3.20)

This equation contains three unknowns and they are (a) the magnitude of R, (b) the phase θ of reflection coefficient, and (c) the value of Δr.

Now to find Δr, we take help of the Figure 3.10 in which $AC = A'C$ and $AD = A'D$ ($\Delta ACD \approx \Delta A'CD$).

From the triangle ABB' (Figure 3.10)

$$r_1 = AB = \sqrt{r^2 + (h_1 - h_2)^2}$$

$$= r\left[1 + \left(\frac{h_1 - h_2}{r}\right)^2\right]^{\frac{1}{2}}$$

$$\approx r + \frac{(h_1 - h_2)^2}{2r}$$

Similarly, from the triangle ABB'

$$r_2 = A'B = \sqrt{r^2 + (h_1 + h_2)^2}$$

$$\approx r + \frac{(h_1 + h_2)^2}{2r}$$

Therefore, the path difference

$$\Delta r = r_2 - r_1 = \frac{1}{2r}\{(h_1 + h_2)^2 - (h_1 - h_2)^2\}$$

$$= \frac{4h_1 h_2}{2r} = \frac{2h_1 h_2}{r}$$

(3.21)

Putting this value of Δr in Equations (3.18) and (3.19), we write

$$\vec{E} = \frac{245\sqrt{P_t G_t}}{r}\sqrt{1 + 2R\cos\left(\theta + \frac{2\pi}{\lambda} \times \frac{2h_1 h_2}{r}\right) + R^2}\, e^{j(\omega t - \varphi)}$$

(3.22)

and

$$\vec{E}_{rms} = 173\frac{\sqrt{P_t G_t}}{r}\sqrt{1 + 2R\cos\left(\theta + \frac{2\pi}{\lambda} \times \frac{2h_1 h_2}{r}\right) + R^2}$$

(3.23)

If now we consider $\theta = 180°$ and $R = 1$, that is, the ground is a perfect reflector, then

$$\vec{E} = \frac{245\sqrt{P_t G_t}}{r}\sqrt{2 - 2\cos\left(\frac{2\pi}{\lambda} \times \frac{2h_1 h_2}{r}\right)}\, e^{j(\omega t - \varphi)}$$

$$= \frac{347}{r}\sqrt{P_t G_t}\sqrt{1 - \cos\left(\frac{4\pi h_1 h_2}{r\lambda}\right)}\, e^{j(\omega t - \varphi)}$$

(3.24)

and

$$\vec{E}_{rms} = \frac{245}{r}\sqrt{P_tG_t}\sqrt{1-\cos\left(\frac{4\pi h_1 h_2}{r\lambda}\right)} \qquad (3.25)$$

Hence the attenuation function, by using Equations (3.20) and (3.21), becomes

$$F = \sqrt{1+2R\cos\left(\theta+\frac{4\pi h_1 h_2}{\lambda r}\right)+R^2} \qquad (3.26)$$

Now to know the value of R, it is necessary to know the grazing angle α, type of polarization, and the electric constants of the reflecting surface. So we are in a position to find the grazing angle first and, then, the effect of polarization. To find the grazing angle, we take the triangle $B'BA$ where $\tan\alpha = \frac{h_1+h_2}{r}$. Since α is small, we may also write $\alpha = \frac{h_1+h_2}{r}$. It is to be remembered that the magnitude of R and phase θ of the reflection point are the function of the reflection point, that is, the distance r, because the grazing angle α varies with r. As r varies the value of F passes through a number of maxima and minima. The maxima occurs when $\cos(\theta+\frac{4\pi h_1 h_2}{\lambda r}) = +1$ and corresponding value of $F = 1+R$. Similarly at minima $\cos(\theta+\frac{4\pi h_1 h_2}{\lambda r}) = -1$ and the value of $F = 1-R$. A typical variation of F as a function of r is shown in Figure 3.11. Here the monotonic decrease of the attenuation function after the vertical line is shown in the same Figure 3.11. It is to be noted that the maxima are smooth and look like the tops of sinusoid, but the minima appear as abrupt changes. For most of the practical purposes, at small grazing angles we may take the liberty to put $R = 1$ and

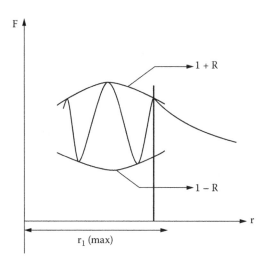

FIGURE 3.11
Attenuation function versus distance (general case).

$\theta = 180°$ (nearly). This is especially true for horizontally polarized wave. Thus the value of F is now becoming in a simplified way (see Equation 3.26)

$$F = 2\left|\sin\frac{2\pi h_1 h_2}{\lambda r}\right|$$

$$(3.27)$$

Hence the distance of maxima can be found from

$$2\pi\frac{h_1 h_2}{\lambda r} = \frac{\pi}{2}(2n+1)\ (n = 0,\ 1,\ 2,\ \text{etc.})$$

Hence

$$r = \frac{4h_1 h_2}{\lambda(2n+1)}$$

and the distance of minima can be found from

$$2\pi\frac{h_1 h_2}{\lambda r} = \pi(1+n)$$

Hence

$$r = \frac{2h_1 h_2}{\lambda(n+1)}.$$

Now, we shall consider the effect of polarization of the transmitted wave. Equations 3.21 and 3.22 hold for horizontal polarization because with horizontal polarization, the direct and reflected wave vectors are in the same direction. But with the vertical polarization, the field vector E_1 wherein the field vector E_1 (Equation 3.22) is lying in the plane of the paper and normal to the line AB. The field vector E_2 (Equation 3.16) is lying in the same plane of paper but normal to the line CB. The angles δ and β made by E_1 and E_2 vectors with the vertical (Figure 3.12) are so small that eventually they may seem to be coincident in direction. If necessary, then we may draw the projections of E_1 and E_2 vectors onto the vertical and horizontal axes, noting that $\tan\delta = (h_1 - h_2)/r$ and $\beta = \alpha$. Thus , we may consider the reflected vector may be horizontally polarized with the incident horizontally polarized wave. The reflected wave would then also be vertically polarized for the incident vertically polarized wave. For example, consider three types of surfaces, namely, sea water ($\epsilon = 80$, $\sigma = 4$ siemens/m); moist soil ($\epsilon = 10$; $\sigma = 0.01$ s/m); and dry soil ($\epsilon = 4$, $\sigma = 10^{-3}$ s/m) from which the reflection occurs. The details are tabulated in Table 3.2. This table shows that the conditions $R = 1$ and $\theta = 180°$ are satisfied with the horizontally polarized wave. For another, the condition $\theta = 180°$ is much easier to satisfy in all cases than the condition $R = 1$. So the criterion for the validity of Equation (3.27) is the condition $R = 1$. So, finally with

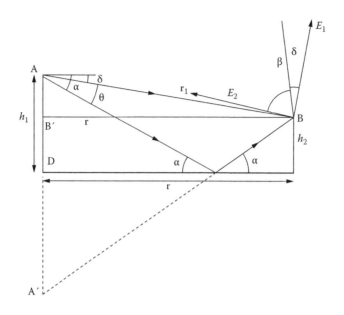

FIGURE 3.12
Same as Figure 3.10 but with vertical polarization.

vertical polarization for the validity of Equation (3.27) decrease of ground conductivity is favorable, whereas with horizontal polarization the same is true of surfaces of highest conductivity. The worst case is radio propagation over sea water at wavelength of 10 m with vertical polarization. But with horizontal polarization, Equation (3.27) may be used practically for all cases.

It is to be remembered that in our discussions we have considered the superposition of two rays, namely, the direct ray and the reflected ray meeting at the receiving end. So Equation (3.27), that is, the attenuation function F, is obtained due to the interference effect, the details of which were discussed earlier.

Moreover, in general, the maxima of F, the attenuation function, will have two values (see Equation 3.27), whereas the minima of F is zero. This is

TABLE 3.2
Finding Maximum Values of $(h_1 + h_2)/r$ for $R = 1$ and $\theta = 180°$

Wave Length (λ)	Maximum Value of $(h_1 + h_2)/r$						Polarization
	Sea Water		Moist Water		Dry Soil		
	$R = 1$	$\theta = 180°$	$R = 1$	$\theta = 180°$	$R = 1$	$\theta = 180°$	
10 m	8×10^{-4}	10^{-3}	8×10^{-3}	6×10^{-2}	10^{-2}	0.25	Vertical
1 cm	3×10^{-3}	0.1	8×10^{-3}	0.3	10^{-2}	0.25	Vertical
10 m	Any	Any	0.08	Any	0.045	Any	Horizontal
1 cm	0.25	Any	0.08	Any	0.045	Any	Horizontal

because we have considered the reflection coefficient is equal to one ($R = 1$). We see again, for further increase of r after the occurrence of the last maxima, the value of F will monotonically decrease with r, as seen in Figure 3.11. On the other hand, the phenomena of interference between the direct wave and reflected wave produce a polarization sensitive lobe structure above the reflecting earth for different angles of θ. However, the reflection of a surface depends on three factors:

1. The reflection coefficient of a plane and smooth surface, ρ
2. The divergence factor
3. Surface roughness $\mathfrak{R}$

So the magnitude of reflection coefficient is $|R| = \rho D \mathfrak{R}$. Here $\mathfrak{R}$ lies between 1 and 0. It depends on whether the surface is perfectly smooth or rough. The divergence factor (D) lies between 1 and 0; for flat earth $D = 1$. The value of D approaches zero as the grazing angle γ approaches zero.

3.6 Effect of the Earth's Curvature: Spherical Earth

If the transmitting and the receiving aerials are a long distance apart, then the spherical nature of the earth has to be considered. The propagation of radio waves between the two elevated aerials would be affected by the curvature of the earth. However, the methodology of finding the total field at the receiver will be the same as we have considered in the case of a flat earth. In that case, the total field would be the sum effect of the direct ray and the reflected ray, as shown in Figures 3.10 and 3.11. The path difference, that is, Δr as used in Equation (3.21), would be altered. Secondly, after reflection from a point on the curved surface of the earth, the ray will be diverging more than what we consider in case of flat earth, as a consequence of which the receiver end will receive less power via the reflected ray (Figure 3.13). In Figure 3.13, let C be the point of reflection on the earth's surface through which a tangent $A'B'$ can be drawn. The respective heights of the aerials from $A'B'$ are called the reduced heights of the antenna. Let these heights be h_1' and h_2' instead of actual heights h_1 and h_2.

We assume the angle of elevation above the convex surface of the earth is the angle of elevation above the tangential plane drawn at C. We also assume that there lies practically no divergence between h_1 and h_2. Thus, we write

$$h_1' = h_1 - \Delta h_1$$
$$h_2' = h_2 - \Delta h_2$$

$$\tag{3.28}$$

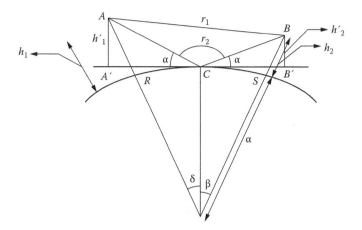

FIGURE 3.13
Path length difference for radio wave propagation over spherical earth.

where $\Delta h_1 = \frac{r_1^2}{2a}$ and $\Delta h_2 = \frac{r_2^2}{2a}$

$$h_1' = h_1 - \frac{r_1^2}{2a}$$

$$h_2' = h_2 - \frac{r_2^2}{2a} \text{ , as } \Delta h \text{ tends to zero} \qquad (3.29)$$

Equation (3.29) shows the reduced heights of the aerials. Now expressing $a = 6.37 \times 10^6$ m and r in kilometers, we write Equation (3.29) as

$$h_1' = h_1(mt) - \frac{r_1^2}{12.8} mt$$

$$h_2' = h_2(mt) - \frac{r_2^2}{12.8} mt \qquad (3.30)$$

Now substituting the values of h_1' and h_2' from Equation (3.30), we get the values of Δr and E_{rms} by using Equations (3.21) and (3.25), respectively.

3.7 Mechanism of Ground Wave Propagation

In this regard, we can have two proximities. First, we may consider that the energy of radio waves is conveyed to a distant point in terms of ray optics. Alternatively, we may say that the energy reaching a point occupies a certain volume in space. To ascertain this let us perform a simple experiment. Let $2x$

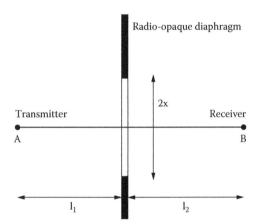

FIGURE 3.14
A simple experiment to find field strength at B.

be the aperture diameter between the transmitter and the receiver, which is kept open to measure the field strength at B (Figure 3.14).

Our objective is to determine (a) the diameter of the volume significant to energy transmission and (b) boundaries of the significant volume, if any.

Now, we gradually shut down the aperture until the field strength at B shows a marked decrease. But no such findings will be noticed. Had the ray optics been applicable then the decrease in field would be noticed if the aperture is fully closed. So at this juncture, instead of applying ray optics, we try to explain the fact with Huygen's principle of wave propagation and the concept of Fresnel zone.

On the basis of Huygen's principle, however, it is possible to determine the received field at any point in space from the known field strength on the surface of a wave front.

According to Figure 3.15, let A be the source and B be the receiving point. The source A is surrounded by a closed surface s in which ds is the elementary

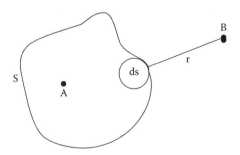

FIGURE 3.15
A is the source and B is the receiving point. The source A is surrounded by a closed surface s in which ds is the elementary area. Here let F be the field at B and F_s be the field at the surface S.

area. Again, we consider F be the field at B and F_s be the field at the surface S. Now obeying Huygen's principle, the field at B is proportional to the field due to the primary wave front at the element and the area of the surface element. Hence,

$$dF = CF_s \frac{e^{-jkr}}{r} ds$$

(3.31)

Here, C is the constant of proportionality. Therefore the total field at B due to the total surface area (considered to be the surface of wave front) is given by Kirchoff's equation as

$$F = -\frac{1}{4\pi} \int \left[F_s \frac{\partial}{\partial n}\left(\frac{e^{-jkr}}{r}\right) - \frac{e^{-jkr}}{r}\frac{\partial F_s}{\partial n}\right] ds$$

(3.32)

where n is the normal directed outward from the plane of the surface S. Knowing the values of F_s and $\frac{\partial F_s}{\partial n}$ one can find out the value of F, that is, the field at B. Now to determine the part of the space that contributes significantly to radio propagation, we draw a sphere (here in two dimensions it should be a circle) of radius l_1, keeping the source A at the center (Figure 3.16).

Now according to Fresnel's idea, we draw the set of curves from B so that they cross the spheres at points $(l_2 + \lambda/2)$. This set of curves forms a conical surface intersecting the plane along BM_1 and BM_1'. Similarly, the higher order of conical surfaces may be drawn at a distance $(l_2 + 2\lambda/2)$ from B, which will intersect along BM_2 and BM_2' and so on. Hence the intersections between the conical surfaces and the sphere will form a system of circles. The segments bounded by adjacent circles are known as Fresnel zones (Figure 3.17).

It is obvious from the wave theory that the waves due to second Fresnel zone are at 180° out of phase with the waves due to first Fresnel zone. This mechanism of phase difference will be maintained in successive zones, which mean that the phase of waves due to the first zone is similar to that of the third zone

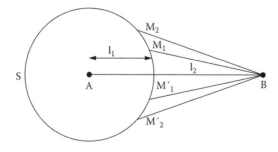

FIGURE 3.16
Determination of the part of the space that contributes significantly to radio propagation, keeping the source A at the center.

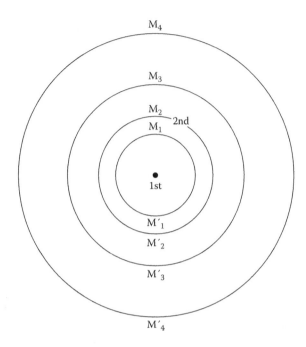

FIGURE 3.17
The segments bounded by adjacent circles are shown as Fresnel zones.

and so on. So, the even-numbered zone will have waves of the same phases and the odd-numbered zones will have phases of the same nature. Hence a pairwise cancellation of the zones will occur. The aggregated effect is thus formed to be equal to about a half of that of the first zone. So, the first Fresnel zone volume will contribute significantly to wave propagation. We are now in a position to calculate the radii of zones. For the nth numbered circle, we write

$$AM_n + M_nB = l_1 + (l_2 + n\lambda/2) \qquad (3.33)$$

To calculate the radii of zones let us draw a schematic diagram (Figure 3.18). Here, M_n is the nth numbered Fresnel zone, which is at a distance x_n above from the line of sight (LOS) path between the source and the receiver. Now referring to Figure 3.18, we write

$$AM_n = \left(l_1^2 + x^2\right)^{1/2} \approx l_1 + \frac{x_n^2}{2l_1}$$

$$\qquad (3.34)$$

$$BM_n = \left(l_2^2 + x^2\right)^{1/2} \approx l_2 + \frac{x_n^2}{2l_2}$$

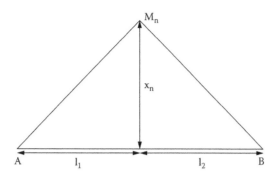

FIGURE 3.18
Schematic diagram of radii of Fresnel zones.

Therefore from Equations (3.33) and (3.34), we write

$$\frac{x_n^2}{2}\left(\frac{1}{l_1}+\frac{1}{l_2}\right)\approx\frac{n\lambda}{2} \tag{3.35}$$

Hence the radius of first Fresnel zone comes out to be

$$x_1=\sqrt{\frac{l_1 l_2 \lambda}{l_1+l_2}} \tag{3.36}$$

It is thus an ellipsoid formed with foci at *A* and *B* whose maximum radius will be at the center of the path.

Hence, the wave propagation through the atmosphere is not like a ray; instead the radio wave is transmitted within a certain volume with a shape of an ellipsoid of revolution bounded by the first Fresnel zone.

References

Dolukhanov, M., 1971, *Propagation of radio waves*, Moscow: Mir Publisher.
Kerr, D. E., 1951, Propagation of short radio waves, Massachusetts Institute of Technology, Radiation Laboratory series, McGraw Hill Book Company, 13, 398.
Long, M. W.,1983, *Radar reflectivity of land and sea*, Norwood, MA: Artech House.
Povejsil, D. J., Raven, R. S., Waterman, P., 1961, *Airborne radar*, Princeton, NJ: D. Van Nostrand.
Shevgaonkar, R. K., 2006, *Electromagnetic waves*, New Delhi: Tata McGraw-Hill.

4

Radio Refraction and Path Delay

4.1 Introduction

Atmospheric refraction comes into play because of the presence of atmospheric irregularities. Due to refraction, the radio wave encounters the variations in atmospheric refractive index along its trajectory that causes the ray path to curve. These types of path distortion while traversing through the atmosphere are very important for many applications like space geodesy. For this purpose, some preliminary discussions have already been made in Section 1.6 of this book. It is to be mentioned that this refraction introduces some delay of the radio path, which is discussed in Section 1.6.5.

4.2 Radius of Curvature of the Ray Path

The atmospheric irregularities due to the variations of meteorological parameters in vertical direction will cause a radio path to bend. Here, we wish to find the curvature of the ray path. In doing so, let us assume, first, the troposphere is consisting of planes parallel to the plane earth (Figure 4.1). We consider two such planes—M and N—separated by a distance dh. A radio ray is incident on the plane M at point a at an angle φ with the normal. Hence the ray would be refracted and will be incident at the point b, say at an angle $\varphi + d\varphi$ on plane N. So, the ray here is turned through an angle $d\varphi$ due to refraction. Evidently, the angle made by ray trace ab at the center of the earth (O) is expressed as $d\varphi$ (Figure 4.1). Therefore, the radius of curvature R is given by

$$R = \frac{ab}{d\varphi} \text{ meter} \tag{4.1}$$

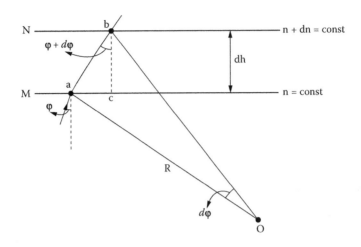

FIGURE 4.1
Determining the radius of curvature of the wave path.

From the triangle $\triangle ABC$, we write

$$ab = \frac{dh}{\cos(\varphi + d\varphi)}$$

$$= \frac{dh}{\cos\varphi\cos d\varphi - \sin\varphi\sin d\varphi}$$

(4.2)

Since $d\varphi$ is small, we neglect the sine component in the expression and hence we are left with

$$ab = \frac{dh}{\cos\varphi}$$

(4.3)

From Equations (4.1) and (4.3), we write

$$R = \frac{dh}{\cos\varphi d\varphi}$$

(4.4)

Applying Snell's law in planes M and N,

$$n\sin\varphi = (n + dn)\sin(\varphi + d\varphi)$$

Neglecting the higher order infinitesimals,

$$n \sin \varphi = n \sin \varphi + n \cos \varphi \, d\varphi + \sin \varphi \, dn$$

or

$$n \cos \varphi \, d\varphi = -\sin \varphi \, dn$$

or

$$\cos \varphi \, d\varphi = -\frac{\sin \varphi \, dn}{n} \tag{4.5}$$

Therefore, from Equations (4.4) and (4.5),

$$R = -\frac{n \, dh}{\sin \varphi \, dn}$$

$$= \frac{n}{\sin \varphi \left(-\frac{dn}{dh}\right)} \tag{4.6}$$

For ground wave propagation, the elevation angles are supposed to be small and hence $\sin \varphi \cong 1$. So without any detriment to the accuracy of the calculation, we set $n \cong 1$ and write

$$R = -\frac{1}{dn/dh} = -\frac{10^6}{dN/dh} \tag{4.7}$$

Hence, in the lower troposphere, the radius of the curvature of the ray path is determined by the lapse rate of refractive index with height and not by the absolute value of N. The negative sign in Equation (4.7) implies that the radius of the curvature will be positive, that is, the propagation path will be convex only when refractive index decreases with height. In the standard troposphere for which the gradient dn/dh is fixed, radiowaves propagated at small elevation angles will travel in arcs of a circle whose radius is given by $R = \frac{10^6}{4 \times 10^{-2}} = 2.5 \times 10^7 \, m = 25,000 \, km$.

It is to be noted that the electromagnetic rays in the visible range 4×10^{14} to 7.5×10^{14} will be least affected during propagation through the troposphere. This is because water molecules with a permanent dipole moment and a finite mass fail to follow the alteration of the electromagnetic field, which occurs at high rate in the visible region of the spectrum. But on the other hand, the electromagnetic waves lower than visible frequency, and water vapor and other polar molecules will fully participate in the oscillatory motion and contribute to the change in refractivity. However, the value of R becomes approximately 50,000 km, for optical frequency radio waves.

4.3 Refractivity is Complex and Frequency Dependent

The refractivity N consists of a frequency independent term, N_0, plus various spectra of refractive dispersion $N'(f)$ and absorption $N''(f)$.

$$N = N_0 + N'(f) + jN''(f) \tag{4.8}$$

The imaginary part of Equation (4.8) expresses the attenuation of the radio signal and the real part expresses the delay with reference to vacuum.

From the application point of view, the imaginary part of Equation (4.8) is expressed as specific attenuation α, and real part is expressed as propagation delay β, (with reference to vacuum); that is, $\alpha = 0.1820\,fN''(f)$ dB/km and $\beta = 3.336[N_0 + N'(f)$ ps/km, where f is frequency in GHz and $N_0 = (2.588P + 2.39e) \theta + N_v$, and θ is the potential temperature.

However, the expression representing N as expressed in Equation (1.14) is considered to be accurate to 0.5% for frequencies less than 30 GHz in normal range of temperature, pressure, and relative humidity. In that case, the total refractivity can be written as the sum of the dry term (N_d) and wet term (N_v), and they are given by

$$N_d = 77.6\frac{P}{T} \tag{4.9}$$

and

$$N_v = 3.73 \times 10^5\,\frac{e}{T^2} \tag{4.10}$$

But, in general, obeying Equation (4.8), that is, for $f > 30$ GHz, the absorption and dispersion spectra may be obtained from line-by-line calculations plus various continuum spectra such as N_d (dry air), N_v (water vapor), and N_w (hydrosols) according to Liebe (1985). However, as mentioned in Equation (4.8), the refractive dispersion and absorption terms are depicted as

$$N''(f) = \left[\sum_{i=1}(SF'')_i + N_d''\right] + \left[\sum_i (SF'')_i + N_v''\right] + N_w'' \tag{4.11}$$

$$N'(f) = \left[\sum_{i=1}(SF')_i + N_d'\right] + \left[\sum_i (SF')_i + N_v'\right] + N_w' \tag{4.12}$$

where S is the line strength in kilohertz (KHz), and F' and F'' are the real and imaginary parts of a line shape function in GHz^{-1} and are given by (Rosenkranz 1975)

$$F''(f) = \left(\frac{1}{X} + \frac{1}{Y}\right)\frac{\lambda f}{v_0} - \delta\left[\frac{v_0 - f}{X} + \frac{v_0 + f}{Y}\right]\frac{f}{v_0}$$ (4.13)

and

$$F'(f) = \frac{z - f}{X} + \frac{z + f}{Y} - \frac{2}{v_0} + \delta\left[\frac{1}{X} - \frac{1}{Y}\right]\frac{\gamma f}{v_0}$$ (4.14)

where

$$X = (v_0 - f)^2 + \gamma^2; \quad Y = (v_0 + f)^2 + \gamma^2$$

and

$$z = \frac{(v_0^2 + \gamma^2)}{v_0}$$

Here, v_0 is the line center frequency; γ and δ are the spectroscopic parameters and are given by

$$\left.\begin{array}{l} S = a_1 p\theta^3 \exp[a_2(1-\theta)], \\ \gamma = a_3(p\theta^{(0.8-a_4)} + 1.1e\theta) \end{array}\right\} \quad \text{for} \quad O_2$$

$$\delta = a_5\theta^{a_6} \quad \text{and,}$$

$$\left.\begin{array}{l} S = b_1 e\theta^{3.5} e \times p[b_2(1-\theta)] \\ \gamma = b_3(p\theta^{0.8} + 4.8e\theta) \end{array}\right\} \quad \text{for} \quad H_2O$$

$$\delta = 0$$

Here, θ is given by $\theta = T(\frac{1000}{P})^{0.286}$. The potential temperature (θ) is defined as that temperature at which the air parcel would take up if brought adiabatically, that is, with no change of heat, up to a standard pressure of 1000 mb.

4.3.1 Continuum for Dry Air

To find $N''(f)$ and $N'(f)$ continuum separate attempts have to be made for dry air and wet air contributions. Now using Equations (4.11) through (4.14) along with spectroscopic coefficients a_1 to a_6 and b_1 to b_3 for strength S and width γ and overlap correction δ are listed in the table (See Table 1 given by Liebe 1985), the dry air continuum part is given by Liebe (1985).

$$N_d''(f) = \left[2a_0 \left\{ \gamma_0 \left(1 + \left(\frac{f}{\gamma_0} \right)^2 \right) \left(1 + \left(\frac{f}{60} \right)^2 \right) \right\}^{-1} + a_d p \theta^{1.5} \right] f p \theta^2 \qquad (4.15)$$

and

$$N_d'(f) = a_0 \left[\left\{ 1 + \left(\frac{f}{\gamma_0} \right)^2 \right\}^{-1} - 1 \right] p \theta^2 \qquad (4.16)$$

A width parameter for the Debye spectrum of O_2 is given by Rosenkranz (1975)

$$\gamma_0 = 5.6 \times 10^{-3} (p + 1.1 e_w) \theta^{0.8} \text{ GHz} \qquad (4.17)$$

The continuum coefficients are $a_0 = 3.07 \times 10^{-4}$ (Rosenkranz 1975) and $a_p = 1.40, (1-1.2 f^{1.5} 10^{-5}) 10^{-10}$ (Stone et al. 1984).

4.3.2 Continuum for Water Vapor

The water vapor continuum is derived empirically (Liebe 1989) from fitting experimental data in the case of N_v'' and based on theoretical data in the case of N_v', leading to

$$N_v''(f) = \left[b_f p + b_v e \theta^3 \right] f e \theta^{2.5} \qquad (4.18)$$

and

$$N_v'(f) \cong b_o f^{2.05} e \theta^{2.4} \qquad (4.19)$$

where $b_f = 1.4 \times 10^{-6}$, $b_v = 5.41 \times 10^{-5}$, and $b_o = 6.47 \times 10^{-6}$.

Water vapor continuum absorption has been a major source of uncertainty in predicting millimeter wave attenuation rates, especially in the window ranges (Bholander et al. 1980). Recent laboratory experiments employing a special high humidity spectrometer of $f = 138$ GHz, $RH = 80$–100%, and $p_1 = 0$ – 150 KPa (nitrogen) yielded the following results (Liebe 1985). Equation (4.18) is needed to supplement local line contributions; coefficient b_f is valid only for the selected local base treated with the line shape (Equations 4.13 and 4.14); the strong self-broadening component $b_v e^2$ is nearly unaffected by coefficient b_f and may be adjusted ($\times 0.915$) for air broadening. The coefficient b_o and both exponents in Equation (4.19) were observed by fitting dispersion results of line-by-line calculations of the rotational H_2O spectrum above 1 THz.

4.3.3 Hydrosol Continuum

According to Liebe (1985)

$$N_w''(f) = 4.50 \frac{w}{\epsilon''(1+\eta^2)} \tag{4.20}$$

and,

$$N_w'(f) \cong 2.4 \times 10^{-3} w\epsilon' \tag{4.21}$$

where $\eta = \frac{2+\epsilon'}{\epsilon''}$; and ϵ', ϵ'' are real and imaginary parts of the dielectric constant for water. For frequencies above 300 GHz, the following approximation

$$N_w'(f) \cong 0.55 w f^{-0.1} \theta^{-6} \tag{4.22}$$

can be used instead of Equation (4.20) based on data reported by Simpson et al. (1979).

The dielectric constant of water is calculated with the Debye model reported by Schaerer and Wilheit (1979), which is valid for $f \leq 300$ GHz.

$$\epsilon'' = \frac{\left(185 - \frac{113}{\theta}\right) f\tau}{1 + (f\tau)^2} \tag{4.23}$$

and

$$\epsilon'' = 4.9 + \frac{185 - \frac{113}{\theta}}{1 + (f\tau)^2} \tag{4.24}$$

Haze conditions are related to conversion of vapor to droplet processes and reversible where $\tau = 4.17 \times 10^{-5} \theta \exp(7.13\theta)$ ns.

Haze conditions are related to conversion of vapor to droplet process and reversible (swelling/or shrinking) with relative humidity for values below RH = 100%. The simple growth function of the haze provides an approximation of the conversion up to RH $\cong$ 96% and typical dry air mass loadings of hygroscopic aerosol in the ground level air are below $w_0 \cong 10^{-4}$ g/m³. A practical growth function covering the range RH = 96–100% is lacking. Such function is needed to model the growth of hydrosols to values $w = 10^{-3}$ to 1 g/m³ as observed under various fog conditions. At present the values for w are assumed to be in Equations (4.20) through (4.22) and the contributions added to calculations for Equations (4.11) and (4.12) for saturated (100% RH) air.

Propagation of microwaves (3–30 GHz) through the neutral atmosphere experiences apparent electrical length (L_e), as

$$L_e = \int_L n(s)ds \qquad (4.25)$$

where $n(s)$ is the refractive index at the position s.

As a matter of fact, the difference of L_e and L, that is, $L_e - L = \Delta L$, is called the excess path delay or simply delay.

Now using Equation (1.14), $N = (n - 1) \times 10^{-6}$, we write

$$\Delta L = 10^{-6} \int N \, ds \qquad (4.26)$$

When we observe an extraterrestrial source through the earth's atmosphere, we usually measure either range or differential range and the slant range R, the path computed from the observer-to-source geometry. However, the presence of atmosphere causes the path of the electromagnetic wave to bend, except at zenith, so that the actual signal path is curved. But, according to Fermat's principle, the actual signal path would be such that the travel time of the signal will be minimum. Thus the geometric length of the signal path in the atmosphere is longer than the slant range, $L > R$. But the electric path must be shorter that is,

$$\int nds < \int ndr$$

According to Hopfield (1971),

$$\int nds - \int ndr \approx -(L - R) \qquad (4.27)$$

The quantity on the left-hand side of Equation (4.27) is called the curvature error. It was noticed by Hopfield that the lower the elevation, the greater is the error. For example, for 15° elevation or even greater, the correction is of the order of few millimeters; for elevation angle 1° or so, the correction is of the order of few meters. According to Chao (1971), the variation of curvature error is larger than the error quoted by Hopfield. It is suggested there that modelling the curvature using standard profiles is not useful. Also, at about approximately 15° elevation, the error due to bending of rays assumes to be of the order of centimeters and the electrical path and the line-of-sight (LOS) path (i.e., slant range) may be taken to be equal for all practical purposes.

However, as discussed in Section 1.6.5, the dry delay is 2.3 m in zenith direction. Gardner (1976) has shown that the effects of horizontal refractivity gradients cause departures from the cosecant of elevation angle of the order of less than 10°. This can produce 3 cm (r.m.s) delay error. These errors are small compared to errors involved in estimating the vapor delay from surface measurements. But in comparison the wet delay term is found to be small, which depends on variation of water vapor density at the particular place in question. To achieve a system for the measurement of wet delay with an accuracy of 1 cm or better, the system should have all weather capability. This system should be suitable for mobile applications and be less expensive. But there is no single measurement technique that fulfills all these requirements. The technique that comes under this preview is the use of passive remote sensing with a dual channel radiometric technique. This system will not be operative in rainy conditions but can operate in clear and cloudy conditions. The details of this technique will be discussed in the subsequent chapters (see Chapter 7) of this book.

4.4 Turbulence-Induced Scintillation

The term *scintillation* refers to those fluctuations about the mean level of received signal, which occur continuously to varying degrees. These are distinguishable from fades due to rainfall. They are distinguishable from deep fades due to low angle fading on the basis of their probability density function (pdf) symmetry.

The literal meaning of *scintillation* is "out of sparkling." When studying the radio wave propagation through the atmosphere, we have to consider a new parameter called the refractivity structure parameter (C_n^2). This gives the measure of the irregularities in refractive index of the troposphere and is defined as the root mean square (rms) difference in the refractive index at the two ports kept unit distance apart. The irregularities are produced from large-scale refractive index gradient caused by turbulence. In the upper troposphere, the water vapor pressure is small and hence only temperature gradients make a significant contribution to the refractive index gradients. But in the lower troposphere, the refractive index is more sensitive to humidity. The refractivity fluctuations associated with turbulence in the troposphere can induce random variations in phase and amplitude of radio waves, which we generally call scintillations. These types of random phase fluctuations limit the performance of radio systems. The small-scale irregularities in refractivity also cause scattering of energy. Sometimes this type of scattering can produce a rapid and deep fading on the radio system (Jassal 1989).

4.4.1 Theoretical Model of C_n^2

Tatarskii (1961) treated the refractive index as a random function with stationary first increments and described its statistical properties in terms of structure function as

$$\varphi n(k) = 0.033 C_n^2 k^{-11/3l} \tag{4.28}$$

where φn is the power spectral density, k is the wave number $(=2\pi/\lambda)$, and l is the scale size.

Equation (4.28) is valid only under the assumption that the turbulent fluctuations are isotropic and l is in the region where the turbulence can be considered isotropic and highly damped by molecular viscosity, which we call inertial subrange. Usually, the inertial subrange for the boundary layer extends from the scales of a few millimeters to tens of meters. Unlike at optical wavelength C_n^2 at microwaves is highly dependent on humidity (Sirkis 1971).

Tatarskii (1961) derived the following expression for C_n^2 at microwave frequencies as

$$C_n^2 = a^2 L_0^{4/3} M^2 \tag{4.29}$$

where a is constant $(=2.4)$, L_0 is the outer scale of turbulence, and

$$M = \left(79 \times 10^6 \frac{P}{T^2}\right)\left(1 + 15500 \frac{q}{T}\right)\left[\frac{dT}{dz} + \tau a - \left(\frac{7800}{\left(1 + 15500 \frac{q}{T}\right)}\right)\right]\frac{dq}{dz} \tag{4.30}$$

where P is the pressure in mb, T is the absolute temperature, q is the specific humidity (dimensionless), z is the altitude (m), and τa is the adiabatic temperature gradient $(=9.8 \text{ K/m})$.

The relation between the variance of the logarithm of intensity $(\ln\sigma^2)$ and C_n^2, according to Tatarskii, is given by

$$\ln \sigma^2 = 0.31 C_n^2 k^{7/6} L^{11/6} \tag{4.31}$$

where L is the path length (m).

4.4.2 Estimation of C_n^2

In its report number 718-2, CCIR (1986) suggested the following expression for the indirect evaluation of C_n^2 based on radio wave propagation measurements. The expression is

$$\sigma_x^2 = 42.25 \left(\frac{2\pi}{\lambda}\right)^{7/6} \int_0^L C_n^2(z) z^{5/6} dz \tag{4.32}$$

where σ_x is the standard deviation of the natural algorithm of the received power, λ is the wavelength (m), and L is the total path length in meters.

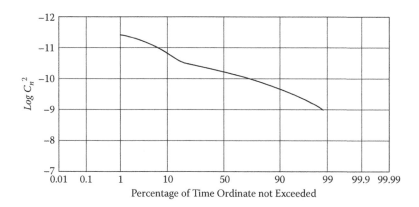

FIGURE 4.2
Cumulative distribution of C_n^2.

Assuming C_n^2 to be constant over the entire path of propagation, Equation (4.32) may be reduced to a simpler form as

$$\sigma_x^2 = 23.04 \left(\frac{2\pi}{\lambda} \right)^{7/6} z^{11/6} C_n^2 \tag{4.33}$$

Jassal (1989) made a propagation experiment using a 157 MHz signal over Roorkee-Derhadun, India. Using Equation (4.33), the values of C_n^2 were computed on an hourly basis from experimentally obtained results.

A cumulative distribution of $\log C_n^2$ for the month of September (Figure 4.2) shows that the distribution follows a lognormal distribution with a standard deviation of 0.7 dB. The C_n^2 values at 99% and 1% probability levels were found to be 8×10^{-11} and 4×10^{-12}, respectively, with the median value 6×10^{-11}, $m^{-2/3}$.

Figure 4.3 shows the diurnal variation of the structure parameter during the month of September. The C_n^2 values for each hour of the day were averaged for 30 days. Hourly variation of C_n^2 values reveals two peaks: one around the midnight and the other around midday. At sunrise, solar heating and the resulting convection may cause an increase in C_n^2 which reaches a peak during afternoon. There was a considerable convection but very little departure from adiabatic temperature gradient. Then the heating decreases during the afternoon and evening periods. As the ground cooled down in the late afternoon and evening periods, the atmospheric stability increases, which causes large gradients. These large gradients require only a small amount of turbulence to cause large C_n^2 values and thus C_n^2 reach a peak around midnight. After midnight, the stability increases, the turbulence dies out, and hence C_n^2 decreases to a minimum, just before sunrise. The cycle continues.

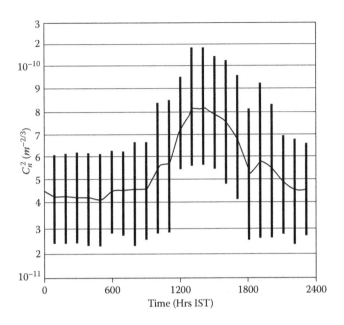

FIGURE 4.3
Diurnal variability of C_n^2.

The vertical bars in Figure 4.3 indicate the overall monthly variation of C_n^2 corresponding to standard deviation. The large values of standard deviation indicate a large spread in C_n^2 values on day-to-day basis. The maximum values of standard deviation were 2.8×10^{-11} during day time. The C_n^2 value as high as $154 \times 10^{-11} m^{-2/3}$ was observed at night during monsoon months. The C_n^2 values as low as 10^{-11} were also found on many occasions during morning and evening hours (Jassal 1989).

When comparing the results of C_n^2 obtained by using the experimentally obtained data, it was found that there exists a large temporal variation in the experimentally obtained results. Van Zandt et al. (1978) have shown that the 4-minute averages of C_n^2 obtained at a height of 10 km can deviate more than an order of magnitude in few tens of minutes. Ecklund et al. (1977) showed that the vertical profile of C_n^2 averaged over a period of 36 minutes decreased by more than two orders of magnitude between 2 and 8 km. Because such a decrease is typical, Gage and Balsley (1978) made a measurement around a circle 13 km in diameter at an altitude of 7 km. Experimental results show that the variation still exists for one order of magnitude from one part of the circle to the other part. Chadvick et al. (1978) made the observations with 160 m range and 1 minute time resolution and showed that at midnight only 5% of the observed C_n^2 values are less than 8×10^{-17} m$^{-2/3}$. The lowest observed values were consistently above 10^{-17} m$^{-2/3}$, and the median value was of the order of 10^{-15} m$^{-2/3}$.

4.5 Microwave Propagation through Tropospheric Turbulence

Microwave propagation experiments through the troposphere reveal the effects of refraction and absorption fluctuations on intensity and phase difference spectra and variances, probability density of intensities and phase difference of the coherence of waves of different wavelength propagated simultaneously, the effect of outer scale of turbulence on intensity statistics, and the refractivity structure constant from intensity scintillations.

In this regard, the parabolic equation, namely, the split-step algorithm developed by Tappert (1977), has become a popular method for studying wave propagation in inhomogeneous media. The method used a paraxial approximation to the time harmonic wave equation to provide a marching solution and give an initial field distribution. The field is calculated as a function of altitude at an arbitrary range (Rouseff 1992). Nowadays it has been used to evaluate electromagnetic wave propagation, although primarily it was used in acoustics and optics (Ko et al. 1983; Dockery 1988; Kuttler and Dockery 1991). The technique is efficient for computational purposes. The field can be calculated for an arbitrary antenna pattern and for an index of refraction varying in both altitude and range. But for microwave propagation the latter point is crucial because surface based and elevated ducts can trap microwave signal. This can significantly affect the coverage pattern of an antenna.

Tropospheric ducts often evolve slowly with time and vary on a large scale compared to microwave wavelength. These ducts, however, are not alone while determining the index of refraction. Superimposed on this large scale and random features are small-scale fluctuations due to tropospheric turbulence. These perturbations may produce cumulative effect on the wave front at a distant range. This is particularly true at observation points where destructive interference from multipaths within the duct might yield a null in the field pattern. It is therefore necessary to introduce the turbulent fluctuations in the propagation model.

Turbulence is often represented as a cascade process where large-scale eddies are broken into successively smaller and smaller sizes until eventually dissipated in the form of heat. The outer scale of the turbulence in the free atmosphere typically ranges from 10 to 100 m. The inner scale is of the order of millimeters and centimeters. Eddies larger than 100 m feed energy into the turbulence. Between these two extremes, the eddies are said to be in the inertial subrange when the kinetic energy of eddies dominates the dissipative effects. Finally, structures less than few millimeters are quickly dissipated due to viscous effect of the atmosphere.

After propagating a short distance, the wave front will deviate from its expected values due to random refractivity fluctuations. Thus total field can then be interpreted as the original coherent wave front plus the small random incoherent component. Then at this situation, the field is said to be weakly scattered. But at greater ranges, the cumulative effect of the random scattering will produce fluctuations that are on the same order as the original signal. Finally, at greater range, the random component dominates and the field is said to be largely incoherent and is strongly scattered.

4.6 Propagation over Inhomogeneous Surface

Radio waves while propagating from transmitter to receiver travel over surfaces of widely different electrical properties. The type of propagation is usually called mixed-path propagation. For simplicity, we consider only two sections that are assumed to be individually uniform. But they differ from one another by the sharply defined (Figure 4.4) boundaries between them. Let us consider, the first part of the propagation path, (r_1) possesses the electrical properties ε_1 and σ_1 and those of the second portion, (r_2) are ε_2 and σ_2. We consider the transmitting aerial is placed at A, whose radiated power and gain are denoted by P_1 and G_1 respectively. Our target is to find the attenuation function at C.

For this purpose, several approaches were put forward since 1930. According to Dolukhanov (1971), the radio waves are assumed to be propagated over particular types of surfaces and hence are attenuated in proportion to the distance over that surface but the attenuation is independent of the location of the surface along the path. In other words it can be stated that if the medium in which radio waves are propagated is linear and isotropic, the attenuation function will be independent of the direction of propagation, which is the reciprocity theorem applied to radio wave propagation.

Shuleikin-Van der Pol's idea (refer to Equation 3.19) was used to determine the actual value of attenuation function F_1 for the path $r_1(F_{1r_1})$ at B. Then we may find out the attenuation function at the same point B on the assumption that the section AB has the properties of the other type of surface, which we call F_2 for the path $r_2(F_{2r_2})$. Finally, the attenuation function at C was found out on the assumption that the entire path has the properties of the second type of surface and we call it as F_3, for the path $(r_1 \neq r_2)$. As is shown in Figure 4.5, the attenuation function is plotted against the distance over the first type of surface, and the broken line is against the distance on the assumption that the entire path has the properties of the other type of surface. The attenuation

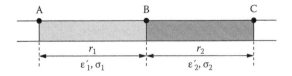

FIGURE 4.4
Mixed-path propagation of a radio wave.

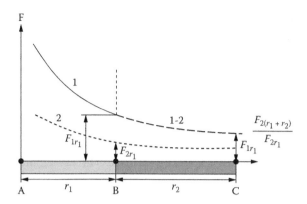

FIGURE 4.5
Attenuation function versus distance for mixed-path propagation.

function at B is F_1/F_2 times its value given by the broken line. Hence, it is assumed that attenuation function at C is as many times its value given by the broken line. So, the actual value of attenuation function at C is given by

$$F_c = \frac{F_3 F_{1r_1}}{F_{2r_2}} \tag{4.34}$$

The adjusted value of attenuation function is shown by the dash-dot line in Figure 4.5. Curve 1 represents the propagation over the first type of surface; curve 2 represents the propagation over the second; and curve 1-2 represents the adjusted attenuation function over the second section of the mixed path.

But, according to the reciprocity theorem, the attenuation function at C will have the same value as it has at A if the transmitter has been shifted to C and the receiver is at A. Then, similar to Equation (4.34), we write

$$F_A = \frac{F_3' F_{2r_2}}{F_{1r_1}} \tag{4.35}$$

where F_3' is the attenuation function at C on the assumption that the entire path has the properties of the first type of surface.

It has been established that the difference in value between F_c and F_A increases as the wavelength decreases and also with the increase in conductivity difference between the two types of surfaces. To over this difficulty, it is proposed (Dolukhanov 1971) that the attenuation function should be the geometrical mean of F_c and F_A and, hence, is written as

$$F = \sqrt{F_c F_A} = \sqrt{\frac{F_3 F_3' F_{1r_1} F_{2r_2}}{F_{1r_2} F_{1r_2}}} \tag{4.36}$$

This method is found to be more satisfactory for all practical purposes.

4.7 Tropospheric Ducting

Tropospheric ducting happens during super refraction. In fact, due to the super refraction of the radio wave it gets trapped in the specified volume of the troposphere. Figures 4.6 and 1.13 will make the idea clear.

We consider point A over the surface of the earth where the transmitting aerial is placed. The rays coming out from the aerial with large elevation angles will suffer only partial refraction and goes to the upper atmosphere (labeled a and b) without being trapped there. The ray emanating from the aerial for which the angle of emergence becomes critical would then be horizontal to the earth at a certain altitude h (labeled c) where $\frac{dN}{dh} = -0.157 \mathrm{m}^{-1}$. The ray for which the emergence angle is more than the critical angle would then suffer total internal reflection and remain at an altitude smaller than h_0 (labeled d). After this the ray would be reflected from the earth's surface and finally traces out the same path as before. This way the ray would remain in the super refractive region in a horizontal direction and produces an abnormally large range beyond the line of sight. This type of propagation phenomena is termed propagation under duct mode. This phenomena may be compared with the propagation in dielectric waveguide. The imperfectly conducting surface of the earth acts as the bottom wall of the waveguide and the upper boundary of the super refractive region as the top wall of the waveguide. As in case of a waveguide, the refractive index inside the super refraction region has a value larger than above it. The difference is that in the waveguide the individual rays experience total internal reflection from both the top and bottom walls; but, on the other hand, in a super refractive region, they undergo ordinary reflection from the surface of the earth and

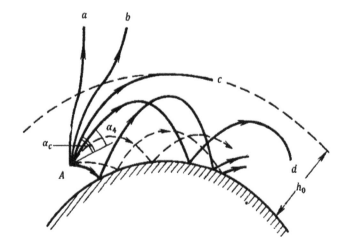

FIGURE 4.6
Radio propagation in case of super-refraction.

total internal reflection from the super refracting volume whose altitude is different from the different angle of elevation.

Thus, the super-refracting region in the troposphere is often referred to as the tropospheric waveguide and the corresponding propagation under the super refraction condition is called ducting. However, the tropospheric duct almost never exceeds 200 m from the surface. Keeping an analogy with the waveguide, it is found that the wavelength of the wave under duct mode does not exceed the critical wavelength λ_c. The relationship between λ_c and the height of the duct is given by, $\lambda_c = 8.5\ h^{3/2} \times 10^{-4}$ m. Usually, the duct mode propagation supports the rays in the range of ultrahigh frequency (300–3000 MHz) and microwaves (3–30 GHz) and millimetric bands also. Ducting has no effect on short, medium, and long waves whatsoever. Although this type of duct mode propagation is an extremely random occurrence, it still occurs uniformly over the sea and happens to occur at certain hours of the day on a fairly regular basis.

4.8 Propagation Delay through the Atmosphere

The radio signals while propagating through the atmosphere suffer refraction. It is known that the refractive characteristics of the neutral atmosphere depend mainly on ambient temperature and water vapor. Since water vapor is a polar molecule, the dipole moment is induced when microwave propagates. Hence the water vapor molecule reorients itself according to the polarity of propagation and causes a change in refractive index of the atmosphere. This refraction introduces uncertainties in the time of arrival due to excess path length or delay(cm) or range error ΔR, defined as (Karmakar et al. 1998; Karmakar et al. 2001)

$$\Delta R = 10^{-6} \int_0^h N(h)\,dh \tag{4.37}$$

where N is defined as refractivity and is given by

$$N = (n - 1)10^6 \tag{4.38}$$

Here, n is the refractive index and is a function of pressure, temperature, and water vapor pressure.

Because the neutral atmosphere is a nondispersive medium, the tropospheric range error estimation is only possible by using models based on easily available atmospheric parameters (Maiti et al. 2009). The propagation delay has two main constituents. These are dry path delay and wet path delay. The dry delay mainly depends on the amount of air through which the signal propagates.

Hence it can be easily modeled with surface pressure measurements. The wet path delay depends on the precipitable water vapor in the column of air through which the signal propagates. Several methods for the estimation of wet path delay have been suggested by Rocken et al. (1991). Among these, path delay or time delay have been estimated directly as a parameter derivable from the refractivity estimation over some places of choice ranging from 22° N to 34° N of Indian subcontinent. Here, in this context, three cities in India were chosen: Kolkata (22° N) with moderate humidity; Jodhpur (26° N), a very dry location; and Srinagar (34° N), with very less humidity.

The term *delay* refers to change in path length due to change in refractive index during the propagation of radio signals through the atmosphere duly constituted by several gases. Their combined refractive index is slightly greater than unity, which gives rise to a decrease in signal velocity. This eventually increases the time taken for the signal to reach the receiving antenna (Adegoke and Onasanya 2008). Besides this, bending of ray path also increases the delay (Collins and Langley 1998).

The refractivity as discussed in Equation (4.38) is divided into two parts. The term N_h is the refractivity due to gases of air, except water vapor, and called the hydrostatic refractivity. The term N_w is the refractivity due to water vapor and is called wet refractivity. Hence from Equation (4.37), we rewrite

$$\Delta R = 10^{-6} \int N_h \, dh + 10^{-6} \int N_w \, dh$$

$$(4.39)$$

The hydrostatic refractivity (N_h) and the wet refractivity (N_w) due to water vapor are respectively equal to

$$N_h = k_1 \, [P_d/T] \qquad (4.40)$$

and

$$N_w = k_2 \, [P_w/T] + k_3 \, [P_w/T^2] \qquad (4.41)$$

where P_d is the partial pressure due to gases, P_w is the water vapor pressure, and T is the ambient atmospheric temperature (Kelvin).

The best average rather than best available coefficients provides a certain robustness against unmodeled systematic errors and increase the reliability of K values, particularly if data from different laboratories can be averaged. However, the best available coefficients according to Rueger (2002) are given by

$$K_1 = 77.674 \pm 0.013 \text{ k/hpa}$$

$$K_2 = 71.97 \pm 10.5 \text{ k/hpa}$$

$$K_3 = 375406 \pm 3000 \text{ K}^2/\text{hPa}$$

Hence, the total refractivity N_r (ppm) is given by

$$N_r = 77.67\ P_d/T + 71.97\ P_w/T + 375406\ P_w/T^2 \qquad (4.42)$$

Because India is a tropical country with a variety of climatic conditions, ranging between 8° N to 36° N, the study of hydrostatic delay using the radiosonde data would be an addition to the studies on navigational aids. The contribution to dry components of atmospheric refractivity comes from an altitude extending from the surface to the lower stratosphere. The prediction of its variability, although spatial and temporal, is found to be easier in comparison to wet components of atmospheric refractivity, which show significant temporal and spatial variability depending on variations in atmospheric water vapor.

The first term in Equation (4.42) shows that partial pressure due to dry gases and the corresponding temperature play a major role in determining dry refractivity. In this respect, three major cities were chosen for which variation of pressure and temperature with height is shown in Figure 4.7 and Figure 4.8, respectively, on June 6, 2005, during the monsoon months. During the monsoon months, the radiometric measurements at Kolkata reveal a value of 60 Kg/m² water vapor content (Karmakar et al. 1999). This happens also during the first week of June. This is the reason for selecting the last day of the first week of June for furthering the study. From Figure 4.8, it is revealed that over Srinagar the ground temperature is substantially low

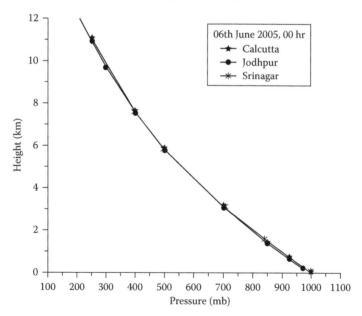

FIGURE 4.7
Variation of pressure with height at three locations.

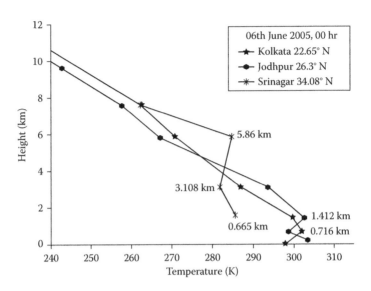

FIGURE 4.8
Variation of temperature with height at three locations.

and the temperature linearly decreases with a height up to 3.1 km and rises again linearly to a height of 5.8 km and again decreases with height. So, over Srinagar temperature inversion occurs twice: one at around 3 km and the other around 6 km. But over Jodhpur, the inversion occurs at around 0.7 km and 1.4 km. But the inversion over Kolkata happens only at around 0.7 km. From Figure 4.8, it appears that the thickness of the inversion layer is large over Srinagar and it is about 3 km. This large thickness in Srinagar in comparison to the other two places is due to the fact that at midnight the hills in Srinagar aid in forming large thickness of radiative cooling of the surface drains from the slopes and uplands into adjacent areas. Because of air drainage, freezing temperatures occur more frequently on bottomlands than on nearby hillsides (Trewartha and Horn 1980). In addition, during the month of June Srinagar is mostly covered with snow. This snow-covered surface gets heated very little due to reflection of solar energy. Moreover, it retards the upward flow of heat from the ground at night. But, on the other hand, Jodhpur being a desert area, contains a porous sandy soil with low heat conductivity, which lowers the temperature of the air layer next to the ground.

We know that lapse rate is defined as the rate of decrease in temperature with height. A temperature inversion means a negative lapse rate, which is pictorially presented in Figures 4.9, 4.10, and 4.11 at three locations, respectively. It is seen from all three cases that there occurs a negative lapse rate over Srinagar at around 5.5 K/km and those over Jodhpur and Kolkata are 43 K/km and 1100 K/km, respectively. So it may be concluded that Kolkata

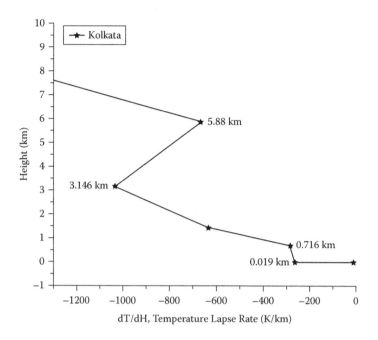

FIGURE 4.9
Variation of temperature lapse rate with height at Kolkata.

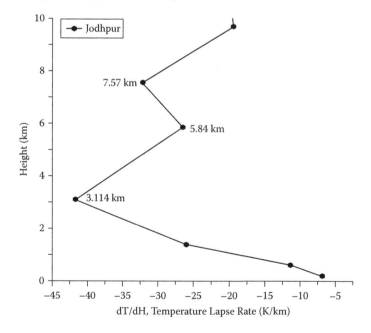

FIGURE 4.10
Variation of temperature lapse rate with height at Jodhpur.

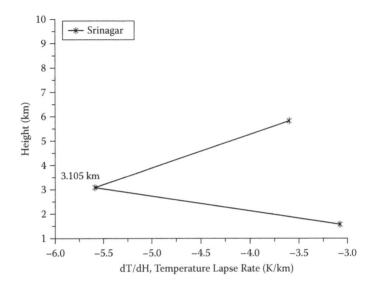

FIGURE 4.11
Variation of temperature lapse rate with height at Srinagar.

being mostly covered with lands and buildings, the ground behaves as a better radiator of heat than air above it.

Hence, the land cools first and then through conduction and radiative exchanges between the lower layers of air and land, the air nearest the ground becomes coldest. This rate of decrease of temperature with increasing height is generally called dry adiabatic lapse rate. But as the air parcel of unsaturated air rises beyond 3.1 Km over Srinagar and cools adiabatically, it will become saturated, that is, the relative humidity becomes 100%.

If the air rises further, condensation begins and cloud forms. This is again verified by exploring the rate of water vapor density variation with height (Figure 4.12).

But the conditions of variations of water vapor density (ρ) variation with height over Jodhpur and Kolkata were entirely different. The analytical formulation of water vapor density were found to be

$\rho = 29.43 \exp(-h/2.05)$ over Kolkata

$\rho = 18.90 \exp(-h/1.77)$ over Jodhpur

$\rho = 15.33 \exp(-h/4.08)$ over Srinagar

The value of $d\rho/dH$ seems to be faster over Kolkata than that over Jodhpur. This value of $d\rho/dH$ over Jodhpur, being a desert area, takes a zero value at about 5.8 Km. According to the same argument favoring the formation of cloud, it may be cited that the formation of cloud over Jodhpur starts at about 5.8 Km and that over Kolkata it starts at about 3.1 Km. Beyond these heights

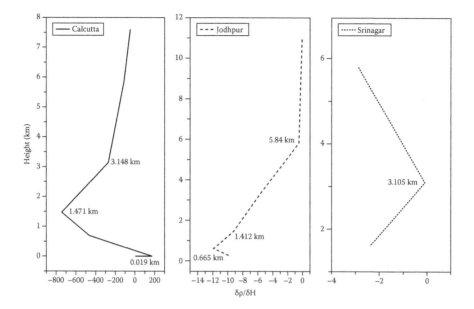

FIGURE 4.12
Variation of water vapor density with height at three locations.

some of water vapor condenses into liquid water droplets in the cloud and corresponding latent heat is released to the air. Nevertheless, the cooling due to expansion of rising air is greater than the latent heating so that the temperature of air continues to decrease but at a slower rate, which is called adiabatic wet lapse rate. This is because the saturated warm air contains much more water vapor than saturated cold water.

4.8.1 Estimation of Refractivity

Following Equation (4.42), it is found that dry refractivity follows the combination of partial pressure and temperature. As dry pressure evolves mainly due to the molecules other than water vapor, it is highly stable and subsequently the dry refractivity also shows a highly stable configuration. To exemplify the situation it was found that over Kolkata the dry refractivity varies between 100 and 250 ppm from the respective height of 11 Km up to ground (Figure 4.12), depending upon pressure and temperature profiles.

On the other hand, the wet refractivity over Kolkata shows a change of slope at around 3.1 km (Figure 4.13), which conforms with the change in dT/dH versus height (H) and the effect has also been observed in the change of $d\rho/dH$ around 3.1 km. So, it may be concluded that the wet refractivity change with height is predominantly determined by temperature and water vapor profile also. Otherwise, if it is due to temperature there only would have been a change in slope in variation of dry refractivity with height. But

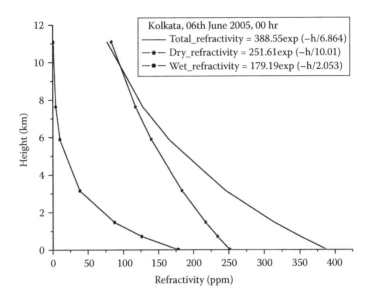

FIGURE 4.13
Variation of refractivity with height at Kolkata.

no such thing happens in this case (Figure 4.13). While summing up the dry and wet refractivity, the same type of slope change occurs as those observed in change of wet refractivity (Figure 4.13). This idea is again supported by observing the variation of dry, wet, and total refractivity with latitude.

It is also observed (Figure 4.14) that dry refractivity does not vary much with latitude, which is considered to be temperature and pressure dependent. But because the water vapor content over Kolkata (22° N) is pretty large, about 60 to 70 kg/m² (Karmakar et al. 1999) in comparison to Jodhpur (26° N) and Srinagar (34° N), the wet refractivity also always assumes a large value (≈177 ppm) over Kolkata than those at high latitudes, whereas wet refractivity over Srinagar is least.

4.8.2 Estimation of Delay (cm)

As discussed in previous sections, dry refractivity depends upon pressure and temperature, but wet refractivity depends predominantly on water vapor concentration in addition to ambient temperature and pressure. To find the variation of delay (cm) in Kolkata, Jodhpur, and Srinagar, Equation (4.38) is used and integrated up to a height of 0 to 10 Km. These delay variations with temperature and water vapor density are presented in Figures 4.16, 4.10, and 4.17. If we compare Figure 4.9 and Figure 4.15, we find that at about 305 K at Kolkata, the inversion occurs near 0.7 Km. It is interesting to note that although the water vapor density profile is almost a smooth exponential, the dry and wet delay variation with temperature and water vapor

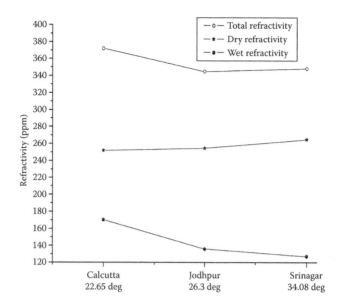

FIGURE 4.14
Variation of refractivity with latitude at three locations.

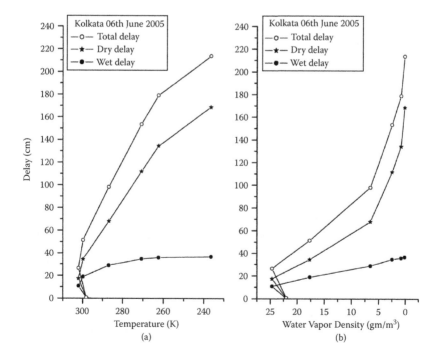

FIGURE 4.15
Variation of delay with temperature (a) and water vapor density (b) at Kolkata.

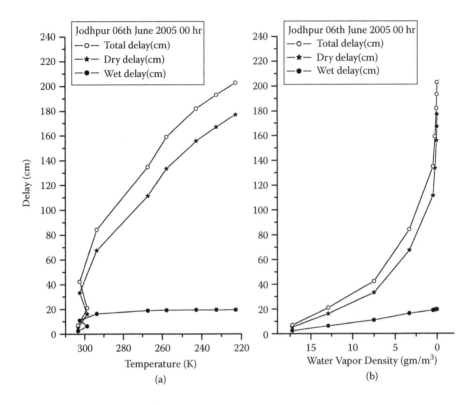

FIGURE 4.16
Variation of delay with temperature (a) and water vapor density (b) at Jodhpur.

density show a prominent inversion at around 0.7 Km at Kolkata. Hence, we may conclude that at Kolkata the delays are mostly affected by temperature, although water vapor may play a major role in determining the integrated value of delay. But on the other hand, if we compare Figure 4.9 and Figure 4.16 for Jodhpur, no inversion occurs there due to change in water vapor density and hence the delay term is predominantly independent on variation of water vapor density.

Similarly, over Srinagar (Figure 4.17) where the wet refractivity is minimum, we find the inversion of delay occurs when the water vapor density becomes 6 gm/m^3. This happens eventually at low altitude of about 3 km or less where cloud liquid may play a major role in determining the delay over Srinagar. Hence it is concluded that we should include water vapor density and cloud liquid when calculating the delay over the cities. Had the location been humid such as Kolkata, then water vapor plays a major role in determining the delay term. In other words, we may say that it is better to look at the variation of water vapor density with height ($d\rho/dH$) in addition to the temperature lapse rate (dT/dH). It is interesting to note that over the humid

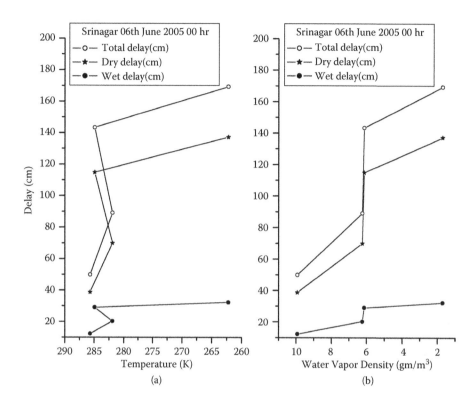

FIGURE 4.17
Variation of delay with temperature (a) and water vapor density (b) at Srinagar.

location (Kolkata) and hilly areas (Srinagar), the cloud formation starts at a lower height where the delay is predominantly determined by temperature and water vapor density and cloud liquid. But at a location where the cloud resides at relatively higher height, that is, at a dry location (Jodhpur), the delay term is mostly dependent on temperature.

The expressions representing the refractivity and hence the delay term are seen to be sufficiently good for the places where the humidity level is less, but for humid locations and hilly areas it is suggested that the inclusion of water vapor and cloud liquid are quite necessary. However, the phase delay induced by cloud droplets can also be approximated using the calculations based on permittivity. Detailed discussions about liquid water attenuation will be made in subsequent chapters. Besides these, the presence of water vapor also produces some absorption of microwave signal propagating through the atmosphere (Adegoke and Onasanya 2008). But this type of absorption is insignificant in low frequencies except over long paths.

References

Adegoke, A. S., Onasanya, M. A., 2008, Effect of propagation delay on signal transmission, *Pacific Journal of Science and Technology*, 9, 1, 13–19.

Bholander, R. A., Emery, R. J., Llewellyn-Jones, D. T., Gimmestad, H. A., Carlon, H. R., Harden, C. S., 1980, Mass spectroscopy of ion-induced water clusters: An explanation of the infrared continuum absorption, *Applied Optics*, 19, 1776–1786.

CCIR, 1986, *Effects of tropospheric refraction on radiowave propagation*, XVIth Plenary Assembly, Study Group V, report 718-2, Dubrovnik, 143–167.

Chadvick, R. B., Moran, K. P., Morrison, G. E., 1978, Measurements towards a C_n^2 climatology, *American Meteorological Society*, 100–103.

Chao, C. C., 1971, Tropospheric range effects due to simulated inhomogeneities by ray tracing, DSN Progress Report 32–1526, 57–66.

Collins, P., Langley, R. B., 1998, Tropospheric propagation delay: How bad can it be? *Proceedings of ION GPS-98, 11th International Tech Meeting of Satellite Division of ION, Nashville, TN*, 15–18.

Dockery, G. D., 1988, Modeling electromagnetic wave propagation in the troposphere using the parabolic equation, *IEEE Transactions on Antennas and Propagation*, AP-36, 1464–1470.

Dolukhanov M., 1971, *Propagation of radio waves*, Moscow: M. R Publisher.

Eckersley, L., 1930, *Proceedings of the IRE*, 18, 1160–1165.

Ecklund, W. L., Carter, D. A., Gage, K. S., 1977, Sounding of lower atmosphere with a portable 50MHz coherent radar, *Journal of Geophysical Research*, 82, 4969–4971.

Gage, K. S., Balsley, B. B., 1978, Doppler radar probing of the clear atmosphere, *Bulletin of American Meteorological Society*, 59, 1074–1093.

Gardner, C. S., 1976, Effects of horizontal refractivity on the accuracy of laser ranging to satellites, *Radio Science*, 11, 1037–1040.

Hopfield, H. S., 1971, Tropospheric effects on electromagnetically measured range: Prediction from surface weather data, *Radio Science*, 6, 357–367.

Jassal, B. S., 1989, *Studies on the tropospheric propagation characteristics of radiowaves over diffraction paths in quasi-mountainous terrains of Northern India* (PhD thesis), Jadavpur University, India.

Karmakar, P. K., Chattopadhyay, S., Sen, A. K., 1999, Estimates of water vapor absorption over Calcutta at 22.235 GHz, *International Journal of Remote Sensing*, 20, 13, 2637–2651.

Karmakar, P. K., Rahaman, M., Chattopadhyay, S., Sen, A. K., Gibbins, C. J., 2001, Estimates of absorption and electrical path length at 22.235 GHz, *Indian Journal of Radio and Space Physics*, 30, 36–42.

Karmakar, P. K., Tarafdar, P. K., Chattopadhyay, S. and Sen, A. K., 1998, Studies on water vapor over a coastal region, *Indian Journal of Physics*, 70 B, 3, 65–71.

Ko, H. W., Sari, J. W., Skura, J. P., 1983, Anomalous microwave propagation through atmospheric ducts, *Johns Hopkins APL Technical Digest*, 4, 12–26.

Kuttler, J. R., Dockery, G. D., 1991, Theoretical description of the parabolic approximation/Fourier split-step method of representing electromagnetic propagation in the troposphere, *Radio Science*, 26, 381–394.

Liebe, H. J., 1985, An atmospheric millimeter wave propagation model, *International Journal of Infrared and Millimeter Wave (USA)*, 10, 6, 631–650.

Liebe, H. J., 1989, An updated model for millimeter wave propagation in moist air, *Radio Science*, 20, 5, 1069–1089.

Maiti, M., Datta, A. K., Karmakar, P. K., 2009, Effect of climatological parameters on propagation delay through the atmosphere, *Pacific Journal of Science and Technology*, 10, 2, 14–20.

Rocken, C., Johnson, J. M., Neilan, R. E., Cerezo, M., Jordon, J. R., Falls, M. J., Ware, R. H., Hayes, M., 1991, The measurement of atmospheric water vapor radiometer comparison and spatial variations, *IEEE Transactions on Geoscience and Remote Sensing*, 29, 1, 3–9.

Rosenkranz, P. W., 1975, Shape of 5 mm oxygen band in the atmosphere, *IEEE Transactions on Antenna and Propagation*, AP-23, 498–506.

Rouseff, D., 1992, Simulated microwave propagation through tropospheric turbulence, *IEEE Transactions on Antennas and Propagation*, 40, 9, 1076–1083.

Rueger, M. J., April 2002, *Refractive index formulae for radio waves*, XXII International Congress, Washington, DC.

Schaerer, G., Wilheit, T. T., 1979, A passive microwave technique for profiling atmospheric water vapor, *Radio Science*, 14, 371–375.

Sen, A. K., Karmakar, P. K., Mitra, A., Devgupta, A. K., Dasgupta, M. K., Calla, O. P. N., Rana, S. S., 1990, Radiometric studies of clear air attenuation and atmospheric water vapor at 22.235 GHz over Calcutta, *Atmospheric Environment (Great Britain)*, 24, 7, 1909–1913.

Simpson, O. A., Bean, B. L., Perkowitz, S., 1979, Far infrared optical constant of liquid water measured with an optically pumped laser, *Journal of the Optical Society of America*, 9, 12, 1723–1726.

Sirkis, M. D., 1971, Contribution of water vapor to index of refraction structure parameter at microwave frequencies, *IEEE Transactions on Antennas and Propagation*, AP-19, 4, 572–573.

Stone, N. W., Read, L. A., Anderson, I., Dagg, R., Smith, W., 1984, Temperature-dependent collision-induced absorption in nitrogen, *Canadian Journal of Physics*, 62, 338–347.

Tappert, F. D., 1977, The parabolic approximation and underwater acoustics, In J. B. Keller, J. S. Papadalos, eds., New York: Springer-Verlag.

Tatarskii, V. I., 1961, *Wave propagation in turbulence medium*, New York: McGraw Hill.

Trewartha, G. T., Horn, L. H., 1980, *An introduction to climate*, New York: McGraw Hill.

Van Zandt, T. E., Green, J. L., Gage, K. S., Clark, W. L., 1978, Vertical profiles of refractivity turbulence structure constant: Comparison of observations by the sunset radar with a new theoretical model, *Radio Science*, 13, 5, 819–829.

5

Absorption of Microwaves

5.1 Introduction

Generally, it is known that solar radiation starts to pass down from the top of the atmosphere and, hence, scattering and absorption take place due to the presence of several molecules in the atmosphere. The abundance of different molecules in the lower troposphere is much larger than those in the upper atmosphere. It was discussed in previous chapters that the presence of water vapor molecules, although small, produces a sizable attenuation in the microwave band.

The absorption exhibited by gaseous molecules, possessing permanent electric and magnetic dipole moment, is due to the coupling of electric dipole and magnetic dipole with electric component and magnetic component of the incoming electromagnetic waves, respectively. This actually causes a change in rotational quantum level producing absorption spectra. However, in this chapter, we shall be dealing with the absorption by several atmospheric constituents mainly predominant in the atmosphere.

In fact, radio waves shorter than 1.5 cm suffer attenuation due to interaction between the field of the wave and the suspended particles of the atmosphere. This absorption is considered in absence of rain, fog, or any other kind of precipitation. The energy of the advancing wave is wasted to heat the gas, or to ionize or to excite atoms and molecules. As they absorb energy, the atoms and molecules pass on from a lower energy state to a higher energy state.

The microwave (3–30 GHz) and millimeter wave (30–300 GHz) interaction with atmospheric gases can be classified into two groups. One of them is resonant interaction and the other is nonresonant interaction. The resonant type is localized to the narrow frequency band, which is supposed to be due to rotational energy level and in some cases the vibration energy levels of the constituent molecules. The nonresonant type of absorption mainly depends upon the bulk properties of the atmospheric constituents as a result of which the wave propagates at a speed less than that in vacuum.

Gaseous absorption in the upper atmosphere removes only about 18% of the incoming energy and 2% absorption approximately are due to the clouds. This is much less than the amount of energy that is absorbed in the

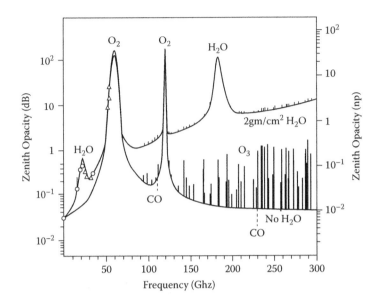

FIGURE 5.1
Atmospheric zenith opacity for H_2O, O_2, O_3, and CO. Also shown are the measured values reported by Staelin (1966), Altshuler and Lamers (1968), and Carter et al. (1968).

troposphere, which is approximately 70%. This is due to the presence of the major part of the ambient atmospheric masses in the troposphere. Such absorption leads directly to the heating effect for which the absorption phenomenon in the troposphere is of great consequence for the energy of the atmosphere. A number of atmospheric molecules have microwave spectral lines out of which the most important lines are due to water vapor and oxygen. Water vapor has rotational lines at 22.235 GHz and 183.311 GHz. Oxygen has a series of rotational spectral lines that combine into a band with a peak at 60 GHz along with a very strong resonant line at 118.75 GHz (Figure 5.1). Besides these, the suspended molecules O_3, CO, N_2O ClO, NH_3, N_2, and CO_2 are present in the atmosphere for which the comprehensive exposition of the microwave spectroscopy have been dealt with elsewhere by several authors. Therefore, their spectroscopy properties are not discussed here in detail. Instead, an effort will be given to find out the absorption experienced by the microwave at water vapor and oxygen resonant lines, and those due to window frequencies, that is, where the microwave spectrum has the minima.

Since the variation of oxygen content is more or less constant in the lower troposphere, the water vapor content variation in the troposphere plays a vital role in presenting absorption spectra, in a particular place of choice. Water vapor content varies from day to day according to the variation of humidity in the ambient atmosphere. Therefore, it is important to know the variation of humidity level to get the idea of water vapor molecular absorption of radio waves. In addition to this, the absolute humidity at which the saturation occurs depends on ambient temperature. Sometimes it happens in the atmosphere that at an elevated temperature the absorption due to the liquid water content (nonprecipitable) of the air becomes more important than that of torrential rain.

5.2 Absorption Coefficient

The absorption coefficient in a medium is a macroscopic parameter. It represents the interaction of the incident electromagnetic energy with the constituent molecules. This is mainly governed by some well-established principles (Janssen 1993):

1. Bohr's frequency condition: The frequency v of a photon emitted or absorbed by the gas is equal to the difference of two energy levels $E_1 - E_2$ of the gas, divided by the Planck's constant.

2. Einstein's law of emission or absorption: If E_1 is higher than E_2 then the probability in the state 1 of emission of a photon from state 1 to state 2 is equal to the probability in state 2 of absorption of photon by transition from 2 to 1 state. These two probabilities are proportional to the incident energy of the radio wave at a particular frequency v. In other words, net absorption is proportional to the difference in thermodynamic probabilities $p_1 - p_2$.

3. Dirac's perturbation theory: For an electromagnetic field, to induce transition between states 1 and 2, the operator with which the field interacts must have a nonzero matrix. For wavelengths that are very long compared to the molecular dimension, this operator turns out to be the dipole moment, which may either be electric or magnetic. Water vapor has an electric dipole moment and therefore the electric field vector of the electromagnetic wave counteracts with the suspended water vapor molecule due to which the absorption by the water vapor molecule takes place. But on the other hand, the oxygen has the magnetic dipole moment, the magnetic component of the electromagnetic field vector counteracts giving rise to absorption due to the oxygen molecules.

5.3 Microwave and Millimeter Wave Absorption in the Atmosphere

The absorption experienced by radio waves is the result of two-fold effects:

1. Absorption by atmospheric gases
2. Scattering by precipitation particles

At wavelengths greater than a few centimeters, absorption by atmospheric gases is generally thought to be negligibly small except where very long distance communications are concerned. However, rain attenuation has to be considered at wavelengths less than 10 cm and is particularly pronounced in the vicinity of 1 cm to 3 cm.

In clear weather, radio waves above 3 GHz experience absorption by water vapor, oxygen, and other gases (such as ozone, hydrogen sulfide, sulfur dioxide, and carbon monoxide). But due to the low molecular densities of ozone (Mitra 1977; Kundu 1983) and carbon monoxide, only water vapor and oxygen contributions are basically significant above the 3 GHz band (Figure 5.1; Staelin 1966; Altshuler and Lamers 1968; Carter et al. 1968). Still, their contributions would be considered in the following sections.

5.3.1 Absorption by Atmospheric Constituents

Absorption exhibited by gaseous molecules, possessing the permanent electric and magnetic dipole moment, is due to the coupling of the electric dipole with the electric field vector and causes a change in rotational quantum level producing the absorption spectra. However, the absorption spectra of these types of molecules, for example, water vapor (asymmetric rotor) and oxygen (linear), are complicated not only by the irregular distribution of energy levels but also because of the complexity of the selection rules and transition probabilities between the levels.

The frequencies at which the electromagnetic field couples to the molecules are determined by the differences between the energy levels of the molecules, according to the equation

$$h\nu = E_1 \sim E_2$$

where h is Planck's constant, ν is frequency, and E_1 and E_2 are the energy levels between which the transition takes place. But, these energy levels are not sharp. It has definite width, which is determined by several effects. These effects are (1) pressure broadening and (2) collision between molecules.

5.3.2 Absorption Intensity

The intensity of a narrow microwave absorption line due to diatomic molecule is given by Townes and Schalow (1975):

$$\gamma = \frac{8\pi^2 N f \, |\mu_{ij}|^2 \, \nu^2 \nabla \nu}{3CkT[(\nu - \nu_0)^2 + (\nabla \nu)^2]} \tag{5.1}$$

where,

N = Number of molecules per cc in the absorption cell

$|\mu_{ij}|^2$ = Square of dipole moment matrix element for the transition, summed over the three perpendicular directions in space

f = Fraction of the molecules in the lower of the two states involved in transition

v = Frequency

v_0 = Resonant frequency, or to a good approximation, the center frequency of the absorption line

Δv = Half width maximum or line width parameter

C = Velocity of light

k = Boltzmann constant

T = Absolute temperature

Neglecting the nuclear spin, f may be written as

$$f = f_{J_{k-1}} k_1 f_v$$

where f_J is the fraction of the molecule in a particular state of rotation J out of f, and $k - 1$ is the rotational state in J for limiting prolate and k_1 is for the limiting oblate asymmetric molecule. Again, f and f_v are given by

$$f_{J_{k-1}k_1} = \frac{(2J+1)\exp(-\frac{W_{J_{k-1}K}}{kT})}{\sum_J (2J+1)\exp(-\frac{W_{J_{k-1}K}}{kT})} \qquad (5.2)$$

$$f_v = \exp\left(\frac{W_v}{kT}\right) \prod_n \left[1 - \exp\left(-\frac{hW_n}{kT}\right) dn\right] \qquad (5.3)$$

where $W_{J_{k-1}}$ is the rotational energy and W_v is the vibration energy. In Equation (5.2) the denominator may be presented by G, where G is called the partition function. At a high temperature $T > 100\ K$, the partition function has an error less than 2% for all known gases.

However, the high temperature approximation may be used for the denominator of Equation (5.2), but it cannot always be assumed that the exponential part in the numerator of Equation (5.2) is approximately equal to unity. For a symmetric molecule, only the lower rotation states give transition in the microwave region, $W_{J_k} \ll kT$, and the Boltzmann factor $\exp(-W_{J_k}/kT)$ may usually be safely set equal to unity. But for asymmetric rotors, however, microwave transition may occur between states each of which has a very large rotational energy, so that the corresponding factor is sometimes considerably smaller than unity and hence it must be retained.

However, Debye assumed the case of the fixed dipole with no rotation or translational energy. After each collision the dipole is assumed not to be random in orientation, but oriented with respect to the electric field present at that moment in accordance with the Boltzmann distribution $-\vec{E}\cdot\vec{\mu}/kT$, where E is the electric field vector and μ is the dipole moment. If the field has oscillated many times before the molecule makes another collision, the dipole has no special orientation with respect to the field at the time of the next collision. During the next collision, however, it is again oriented with respect to the electric field vector of the electromagnetic wave causing absorption. Such a process is repeated many times causing absorption of energy although there is no peak resonance absorption. And accordingly Debye calculated the absorption coefficient as

$$\gamma = \frac{\omega}{c} \frac{4\pi N\mu^2}{3kT} \frac{\omega\tau}{1+\omega^2\tau^2} \tag{5.4}$$

where,
   ω = Angular frequency of the incident radiation
   τ = Mean lifetime between collisions
   N = Number of molecules per cc
   μ = Dipole moment

In fact, this Debye theory and the Lorentz theory of absorption have been synthesized by Van Vleck and Weisskopf with the assumption that the molecules undergo a violent collision and the phase of its oscillation after such a collision will not be greatly dependent on its phase at the start of its collision. Under this assumption the phase after a collision is arbitrary. However, above all, to determine the absorption one needs only to solve the equation of motion of classical linear oscillator in the field subject to the right boundary conditions. The equation is then written as

$$\ddot{x} + \omega_0^2 x = \frac{eE}{m}\cos(\omega t) \tag{5.5}$$

where,
   ω_0 = 2π times the natural molecular frequency
   ω = 2π times the frequency of oscillation of the electric field E

Now, if x has been averaged over all molecules, Equation (5.5) reduces to

$$\bar{x} = aE\cos(\omega t) + bE\sin(\omega t) \tag{5.6}$$

where a and b are the constants and eventually we get them as

$$a = \frac{e}{m(\omega_0^2 - \omega^2)}\left[1 - \frac{\omega}{2\omega_0^2\tau^2}\left\{\frac{\omega_0 + \omega}{(\frac{1}{\tau})^2 + (\omega_0 - \omega)^2} + \frac{\omega_0 - \omega}{(\frac{1}{\tau})^2 + (\omega_0 + \omega)^2}\right\}\right] \tag{5.7}$$

$$b = \frac{e\omega}{2m\omega_0^2\tau}\left[\frac{1}{(\frac{1}{\tau})^2 + (\omega_0 - \omega)^2} + \frac{1}{(\frac{1}{\tau})^2 + (\omega_0 + \omega)^2}\right] \tag{5.8}$$

To obtain the fractional absorption per unit distance, let us consider unit cube within which the radiation is traveling through keeping one of its faces perpendicular to the direction of propagation. The radiation then, absorbed during a time (T) will be

$$n \int_0^T e \, \dot{\overline{x}} E \cos(\omega t) dt$$

The total energy passing into the cube will be $C(E^2/8\pi)T$, so that fractional absorption per unit distance is

$$\gamma = \frac{n \int_0^T e \, \dot{\overline{x}} \, E \cos(\omega t) dt}{C(E^2/8\pi)T}$$

Remembering again the form of $\dot{\overline{x}}$ and integrating over a long time, the absorption coefficient comes to be

$$\gamma = \frac{4\pi neb\omega}{C} \tag{5.9}$$

So, the absorption coefficient from Equation (5.9), on substitution of b we get

$$\gamma = \frac{ne^2\omega^2}{mcv_0^2} \left[\frac{1/2\pi\tau}{(v_0 - v)^2 + (\frac{1}{2\pi\tau})^2} + \frac{1/2\pi\tau}{\left(v_0 + v\right)^2 + (\frac{1}{2\pi\tau})^2} \right] \tag{5.10}$$

But when the quantum mechanical case (Heitler 1954) is being considered the term e^2/m in Equation (5.10) has to be replaced by $(\frac{8\pi^2}{3h}) |\mu_{ij}|^2 \, v_0$ where $|\mu_{ij}|$ is the matrix element of the dipole moment or simply called the dipole moment for the transition from i to j state. This is given by

$$|\mu_{ij}|^2 = \sum |\mu_x(JMJ'M'|^2 + |\mu_y(JMJ'M'|^2 + |\mu_z(JMJ'M'|^2 \tag{5.11}$$

In addition to these two states, there may be many other molecular states that are occupied by the molecules of the material of the cell under consideration. In that case, we may write f, the fraction of the total molecules that is in the lower state in such a way so that n in Equation (5.10) may be replaced by Nf, where N is the total number of molecules per unit volume. Making this substitution we get,

$$\gamma = \frac{8\pi^2 Nf}{3CkT} |\mu_{ij}|^2 \, v_0^2 \left[\frac{1/2\pi v}{(v_0 - v)^2 + (\frac{1}{2\pi v})^2} + \frac{1/2\pi v}{(v_0 + v)^2 + (\frac{1}{2\pi v})^2} \right] \tag{5.12}$$

But according to Anderson et al. (1959), a close approximation of the absorption coefficient similar to that obtained by Van Vleck and Weisskopf can be written as

$$\gamma = \frac{8\pi^2 Nf}{3CkT} |\mu_{ij}|^2 \ v^2 \left[\frac{\Delta v}{(v_0 - v - a\Delta v)^2 + (\Delta v)^2} + \frac{\Delta v}{(v_0 + v + a\Delta v)^2 + (\Delta v)^2} \right] \qquad (5.13)$$

where $a\Delta v$ is the frequency shift, However, in the microwave and millimeter wave region this frequency shift is not so important (Townes and Schalow 1975) and accordingly Equation (5.13) takes the form

$$\gamma = \frac{8\pi^2 Nf}{3CkT} |\mu_{ij}|^2 \ v^2 \left[\frac{\Delta v}{(v_0 - v)^2 + (\Delta v)^2} + \frac{\Delta v}{(v_0 + v)^2 + (\Delta v)^2} \right] \qquad (5.14)$$

where Δv is the half width of the absorption line and other symbols are as stated earlier. For, $v_0 \gg \Delta v$ the second term in the parenthesis has been neglected in Equation (5.14) and we get,

$$\gamma = \frac{8\pi^2 Nf}{3CkT} |\mu_{ij}|^2 \ v^2 \left[\frac{\Delta v}{(v_0 - v)^2 + (\Delta v)^2} \right] \qquad (5.15)$$

Now at $v = v_0$ we have

$$\alpha_0 = \frac{8\pi^2 Nf}{3CkT} |\mu_{ij}|^2 \ \frac{1}{\Delta v} \ cm^{-1} \qquad (5.16)$$

So, in terms of v_0 Equation (5.15) can be written as

$$\gamma = \alpha_0 \frac{\Delta v}{(v_0 - v)^2 + (\Delta v)^2} \ \frac{v^2}{v_0^2} cm^{-1}$$

$$\gamma = \alpha_0 \frac{\Delta v}{(v_0 - v)^2 + (\Delta v)^2} \ \frac{v^2}{v_0^2} \ 10^6 \log_e 10 \ dB/km \qquad (5.17)$$

$$\gamma = \gamma_0 \frac{(\Delta v)^2}{(\gamma_0 - v)^2 + (\Delta v)^2} \ \frac{v^2}{\gamma_0^2} \qquad (5.18)$$

$$\gamma = \gamma_0 \frac{1}{1 + (\frac{\gamma_0 - v}{\Delta v})^2} \ \frac{v^2}{\gamma_0^2} \qquad (5.19)$$

Where, $\gamma_0 = 10^6 \alpha_0 \log_e 10$.

To obtain the value of γ three sets of curves are required (Ghosh and Edwards 1956; Malaviya 1961; Ghosh and Ghosh 1986; Ghosh and Kumar 1983):

1. $\gamma_0 v_0$ with altitude
2. Δv with altitude

$\dfrac{1}{[1+(\frac{v_0-\gamma}{\Delta\gamma})^2]}$. with γ

From Equation (5.15) it follows that for the calculation of microwave and millimeter wave absorption, we require the variation of γ_0 and Δv for each constituent of the atmosphere, with altitude.

It has been found experimentally that for all microwave absorbing gases, the value of Δv varies linearly with pressure over a wide range of low pressure (Gordy et al. 1953). It follows, therefore, from Equations 5.16 and 5.19 that γ_0 is independent of pressure.

According to Morse and Maier (2011) f in Equation (5.14) is rewritten as

$f \propto \frac{1}{T}$, for diatomic and linear polyatomic molecules (e.g., O_2, NO, CO, N_2O)

$f \propto \frac{1}{T^{15}}$, for symmetric and asymmetric tops (e.g., H_2O, H_2S, SO_2, NO_2, O_3).

Moreover, substituting $N = P/kT$, where P is the partial pressure of the absorbing gas, we obtain for O_2, NO, CO, and N_2O

$$\alpha_0 = \frac{8\pi^2 v_0^2 \mu_{ij}^2 P}{const.3CkT^3} \frac{1}{\Delta v} \tag{5.20}$$

and for H_2O, H_2S, SO_2, and so forth,

$$\alpha_0 = \frac{8\pi^2 v_0^2 |\mu_{ij}|^2 P}{const.3CkT^{7/2}} \frac{1}{\Delta v} \tag{5.21}$$

The variation of Δv with temperature and pressure is given by Gordy et al. (1953)

$$\Delta v = (\Delta v)_1 P_{mm} \left(\frac{300}{T}\right)^3$$

where $(\Delta v)_1$ is half width at 300 K and at 1 mm pressure. Substituting this in Equations (5.18) and (5.19) we get,

α_0 proportional to $\dfrac{1}{T^{3-x}}$ for O_2, NO, CO_2 and N_2O

α_0 proportional to $\dfrac{1}{T^{3.5-x}}$, for H_2O, H_2S, SO_2, NO_2 and O_3

The value of x for oxygen has been found to be $\frac{3}{4}$ (Ghosh and Ashoke 1983) and for O_3, H_2O, N_2O, NO, H_2S, SO_2, and CO, the value of x has been assumed to be unity (Gordy et al., 1953).

Since $\gamma_0 = 10^6 \alpha_0 \log_e 10$,   thus

$$\gamma_0 \propto \frac{1}{T^{\frac{9}{4}}} \text{ for } O_2$$

$$\propto \frac{1}{T^2}, \text{ for } N_2O, NO, CO$$

$$\propto \frac{1}{T^{2.5}} \text{ for } H_2O, NO_2, SO_2, H_2S \text{ and } O_3$$

that is, γ_0 is proportional to α_0 which is again proportional to $\frac{1}{T^{3-x}}$ for O_2 and N_2O.

Therefore,

$$\frac{\alpha_0(T)}{\alpha_0(300)} = \left(\frac{300}{T} \right)^{3-x}$$

where $\alpha_0(T)$ and $\alpha_0(300)$ represent α_0 at T K and at 300 K, respectively.

So,

$$\gamma_0(T) = 10^6 \alpha_0(T) \log_e 10$$

$$= 10^6 \alpha_0(300) \left[\frac{300}{T} \right]^{3-x} \log_e 10,$$

or

$$\frac{\gamma_0(T)}{\alpha_0(300)} = 10^6 \left[\frac{300}{T} \right]^{3-x} \log_e 10$$

Similarly for H_2O and O_3

$$\gamma_0(T) \propto \alpha_0(T) \propto \frac{1}{T^{3.5-x}}$$

and hence,

$$\frac{\gamma_0(T)}{\alpha_0(300)} = 10^6 \left[\frac{300}{T} \right]^{3.5-x} \log_e 10$$

Now, variation of γ_0 (T) with altitude can be plotted for

$$T^{3-x} \text{ for } x = 3/4, \text{ for } O_2$$

$$T^{3-x} \text{ for } x = 1, \text{ for other constituents}$$

$$T^{3.5-x} \text{ for } x = 1, \text{ for other constituents}$$

From these curves $\gamma_0(T)$ at any altitude can be obtained (Figure 5.2), provided their absorption coefficients at 300 K and at peak frequencies, are known (Ghosh and Edwards 1956).

Now to obtain the second set of curves we take help of collisional broadening theory of absorption lines (Artman 1953). According to Artman,

$$(\Delta v)_{mixture} = (\Delta v)_1 \left[\frac{300}{T} \right]^x (P_{O_2} + \beta P_{N_2})$$

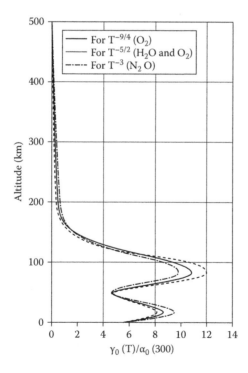

FIGURE 5.2

Variation of γ_0 $(T)/\alpha_0$ with altitude for oxygen and other atmospheric gases for $T^{-9/4}$ (O_2); $T^{-5/2}$ (H_2O and O_3); and for T^{-3} (N_2O).

where P_{O_2} and P_{N_2} are the partial pressure due to oxygen and nitrogen molecules, respectively. Artman found $\beta = 0.75$ and $(\Delta v)_1 = 1.94$ MHz/mm at 300 K. Hence, for oxygen

$$(\Delta v) = 1.9 + \left[\frac{300}{T} \right]^{\frac{3}{4}} (P_{O_2} + 0.75 P_{N_2})$$

For O_3, H_2O, SO_2, N_2O, CO_2, and H_2S value of x is taken to be equal to unity and then,

$$(\Delta v) = (\Delta v)_1 P_{mm} \frac{300}{T}$$

where

$(\Delta v)_1 = 25$ MHz/mm for O_3, NO_2, NO, H_2S, SO_2, CO
$(\Delta v)_1 = 14$ MHz/mm for H_2O
$(\Delta v)_1 = 4.2$ MHz/mm for N_2O (Ghosh and Edwards 1956)

And accordingly, variation of Δv with altitude can be obtained (Figure 5.3). Last, to obtain the last set of curves, a particular value of v is to be chosen and then the values of the expression $\dfrac{1}{\left[1 + \left(\frac{\gamma_0 - \gamma}{\Delta \gamma} \right)^2 \right]}$ are calculated for different values of $v_0 - v$ and then plotted in Figure 5.4.

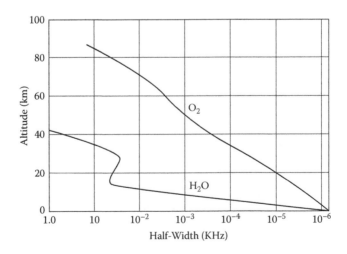

FIGURE 5.3
Variation of $\Delta \gamma$ with altitude for O_2 and H_2O.

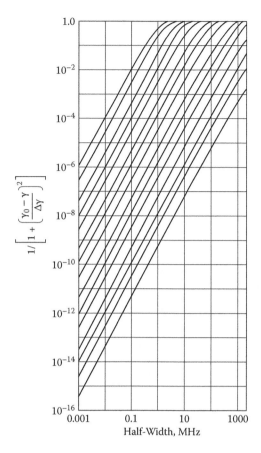

FIGURE 5.4

Variation of $\dfrac{1}{\left[1+\left(\frac{\gamma_0-\gamma}{\Delta\gamma}\right)^2\right]}$ with Half Width at different values of $(\gamma_0 - \gamma)$.

Finally, by using Equation (5.9) the total absorptions of the terrestrial atmosphere can be obtained, which are given in Figure 5.5.

5.4 Centrifugal Distortion

Besides the complexity of the symmetric and asymmetric spectral structure, centrifugal distortion is tremendously more important in microwave spectra of asymmetric rotors than in the spectra of symmetric rotors. In symmetric tops it produces very small shifts of the order of 1 Mc or less, but in some asymmetric rotors, centrifugal distortions causes the change of the rotational

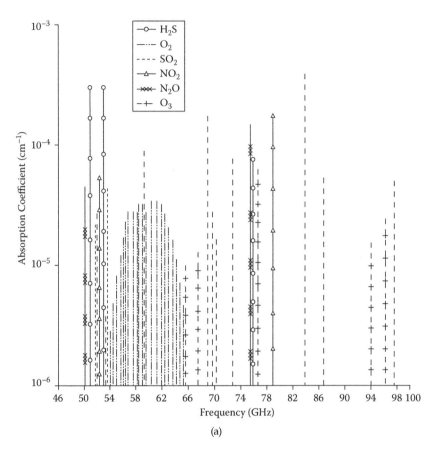

FIGURE 5.5
Calculated absorption coefficient in cm⁻¹ for frequencies 4 to 400 GHz in (a)–(d).

frequencies many hundreds of megacycles (Townes and Schalow 1975). This is because microwave transition of asymmetric rotors may occur between states of rather large angular momentum and of very large rotational energies. On the other hand, in symmetric tops, microwave transition may occur between rather smaller J values. But, if the transition takes place between larger J values, for example, in the case of symmetric rotors, the larger value of moment of inertia makes the rotational energy between the levels concerned still smaller.

To make it clear we take an example of the light molecule H_2O (asymmetric rotor). The transition of H_2O which lies at 22.235 GHz involves levels with rotational energies near 500 cm⁻¹ or 1.5×10^7 Mc, even though J is only 5 or 6 (Townes 1975).

In the case of infrared region, water vapor peak absorption takes place at several frequencies, out of which $J = 11$ may be considered, for which centrifugal distortion correction occur that are as large as 9% of the entire rotational energy or 280 cm⁻¹ (Randall et al. 1937).

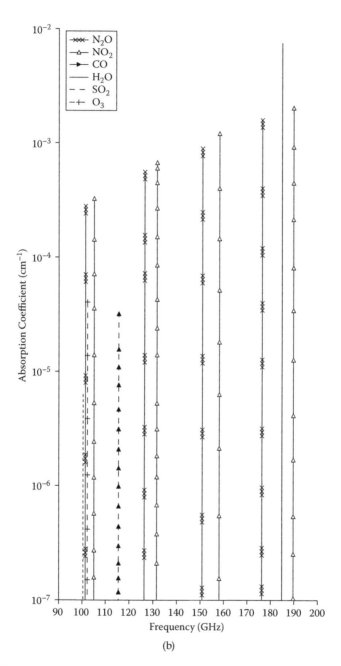

FIGURE 5.5
(Continued)

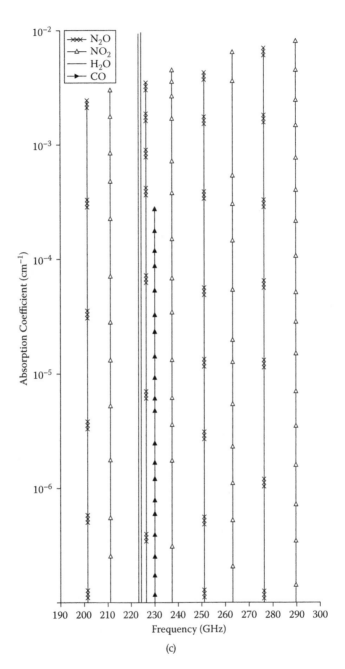

(c)

FIGURE 5.5
(Continued)

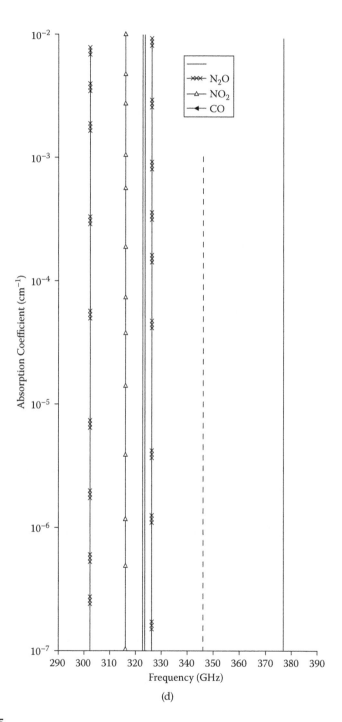

(d)

FIGURE 5.5
(Continued)

5.5 Water Vapor Absorption at 22.235 GHz

In the microwave absorption spectra (Figure 5.1), we have two rotational lines occurring at 22.235 ($\gamma = 1.35$ cm) and the other at 183.311 GHz ($\gamma = 1.64$ mm). These two lines represent the water vapor absorption maxima up to 300 GHz.

The absorption $\gamma = 1.35$ cm can be written as the sum total of absorption due to resonance and that due to nonresonance part (Van Vleck 1947a). It is given by

$$\gamma_{total} = \gamma_{resonance} + \gamma_{nonresonance}$$

where

$\gamma_{resonance}$ = Contribution to the absorption due to resonance at $\gamma = 1.35$ cm
$\gamma_{nonresonance}$ = Contribution to the absorption due to the wings of higher frequencies

Let's now consider water vapor resonance line at $\gamma = 1.35$ cm. Equation (5.14) as derived by Townes and Schalow (1975) using the spectroscopic analyses, can be written as (Bhattacharya 1985)

$$\gamma_{1.35} = \frac{8\pi^2 N |\mu_{ij}| v^2 \exp(-E_{5,-1})/kT}{3CkTG} \left[\frac{\Delta v}{(v_0 - v)^2 + (\Delta v)^2} + \frac{\Delta v}{(v_0 + v)^2 + (\Delta v)^2} \right]$$

$$+ \frac{16\pi^2 N v^2 \Delta v}{3TkCG} \sum |\mu_{ij}|^2 \frac{\exp(-E_j/kT)}{v_{ij}}$$

(5.22)

where v_0 is the resonance frequency and G is the partition function, which can be expressed as (Van Vleck, 1947a)

$$G = \sum_{j=0}^{\infty} \sum_{\tau=-j}^{+j} [2 - ((-1)^\tau)](2J + 1)\exp(-E_i \tau/kT) = 0.34\, T^{3/2}$$

(5.23)

According to Denison (1940), we write $E_{5,-1} = 446.39$ cm^{-1}.
The value of electric dipole moment is given by King et al. (1947) as

$$|\mu_{ij}|^2 = 0.165\mu_0^2$$

where μ_0 is the permanent dipole moment of water vapor, which is given by (Bhattacharya 1985)

$$\mu_0 = 1.84 \times 10^{-18} \text{ e.s.u}$$

and hence

$$|\mu_{1.35}| = 5.6 \times 10^{-37} \text{ e.s.u}$$

Now, using all these constants and also Van Vleck's evaluation of summation in the second term of Equation (5.1), we get the equation as used by Croom (1965)

$$\gamma_{1.35} = 1.05 \times 10^{-28} \frac{Nv^2 \exp(-644/T)}{T^{5/2}} \left[\frac{\Delta v}{(v_0 - v)^2 + (\Delta v)^2} + \frac{\Delta v}{(v_0 + v)^2 + (\Delta v)^2} \right] + 1.52$$

$$\times 10^{-52} Nv^2 \Delta v \Big/ T^{3/2}$$

$$(5.24)$$

where

v = Frequency in Hz
T = Kinetic temperature in K
Δv = Half width parameter, which is given by

$$\Delta v = 2.62 \times 10^9 \frac{P/760}{\left(T/318\right)^{0.625}} (1 + 0.0046\rho) \text{ sec}^{-1} \qquad (5.25)$$

where,

ρ = Water vapor density in gm/m3
P = Total atmospheric pressure in mm/Hg
N = Number of water vapor molecules per cm^3

According to Croom (1965), the nonresonant part of Equation (5.3) is given by

$$\gamma_{nonresonance} = 1.52 \times 10^{-52} \frac{Nv^2 \Delta v}{T^{3/2}}$$

But Gaut (1968) pointed out that this value is not in close agreement with the experimental data and hence is to be multiplied by a factor 5. So the nonresonance term comes out as

$$\gamma_{nonresonance} = 5 \left(1.52 \times 10^{-52} \frac{Nv^2 \Delta v}{T^{3/2}} \right)$$

$$= 2.55 \times 10^{-8} \frac{\rho v^2 \Delta v}{T^{3/2}}$$

This is the expression as used by Falcone et al. (1971) in units of cm^{-1}. But to provide a better agreement with the experimental study of Becker and Autler (1946), Chung (1962) gives the following expression obtained by scaling the expression as given by Van Vleck:

$$\gamma_{resonance} = \frac{4.32 \times 10^{-4} exp\left(-644/T\right)}{T^{3.125}}\left(1 + 0.011\rho T/P\right)$$

$$\times \left[\frac{1}{(\nu_0 - \nu)^2 + (\Delta\nu)^2} + \frac{1}{(\nu_0 + \nu)^2 + (\Delta\nu)^2}\right].$$

and

$$\gamma_{nonresonance} = 2.55 \times 10^{-8}\rho\nu^2 \Delta\nu/T^{3/2}$$

where

$$\Delta\nu = 2.61 \times 10^9 \left(1 + 0.011\rho T/P\right)\frac{P/760}{(T/318)^{0.625}} \tag{5.26}$$

as given by Benedict and Kaplan (1959). Now using the expression

$$\gamma_{total} = \gamma_{resonance} + \gamma_{nonresonance}$$

one can find out the total absorption at $\lambda = 1.35$ cm which is given by

$$\gamma_{1.35} = \frac{4.32 \times 10^{-4}exp(-644/T)P\rho\nu^2}{T^{3.125}}\left(1 + 0.011\rho T/P\right)$$

$$\times \left[\frac{1}{(22.235 - \nu)^2 + (\Delta\nu)^2} + \frac{1}{(22.235 + \nu)^2 + (\Delta\nu)^2}\right] + 2.55 \times 10^{-8}\rho\nu^2 \frac{\Delta\nu}{T^{3/2}}$$

$$\tag{5.27}$$

But this equation is not valid for frequencies higher than approximately 75 GHz because of presence of other resonance lines above 70 GHz.

5.6 Water Vapor Absorption at 183.311 GHz

It has already been discussed in the energy spectrum that the consecutive water vapor lines occur at 22.235 GHz and 183.311 GHz. But the 22.235 GHz line is a very weak line. Theoretical estimates (Ulaby et al. 1986) show that at or near the peak of the water vapor line, the absorption parameter is dominated by the water vapor component. This is true for the strong 183.311 GHz line and also true for the weak 22.235 GHz except in dry climates. According to Smith (1953) the absorption coefficient at 22.235 GHz is larger than absorption exhibited by a factor 20 (for water vapor density $\rho_0 = 7.5$ gm/m). But in dry climate, say, for $\rho_0 = 1$ gm/m3, the ratio comes down to 3 only and the total absorption coefficient is not as strongly dominated by water vapor alone at 22.235 GHz. But attenuation at 183 GHz is found to be roughly 130 times than that measured at 22 GHz and hence 183 GHz line is considered as a very strong line.

The absorption coefficient at 183 GHz, that is, $\lambda = 1.64$ mm, is well discussed by Van Vleck (1947b), whereas Barrett and Chung (1962) gave a simplified expression as

$$\gamma_{1.64mm} = 6.46 \times 10^{-29} \frac{N v^2 e^{-200/T}}{T^{2.5}} \left[\left[\left[\frac{\Delta v}{(v_0 - v)^2 + (\Delta v)^2} + \frac{\Delta v}{(v_0 + v)^2 + (\Delta v)^2} \right] \right] \right]$$

$$+ 1.8 \times 10^{-52} \frac{N v^2}{T^{1.5}} \Delta v$$

(5.28)

where
  N = Number of molecules per cm³
  v = Frequency in Hz
  v_0 = Resonance frequency in Hz
  T = Kinetic temperature in K
  Δv = Line width parameter and is given by Equation (5.25)

This 183.311 GHz line is so strong that it cannot be used for any ground-based observation because it screens off the radiations originating at a very large distance from the receiving antenna. Thus only the lower atmosphere can be well studied by balloon-borne or satellite-borne 183.311 GHz radiometric studies. Moreover, the water vapor weighting function variation with height is very stable. This implies a very good vertical resolution. Hence, 183 GHz line appears to be very useful choice for profiling of water vapor.

This line gives a very excellent method to calculate very small variation of water vapor in the stratosphere also. According to Croom (1965) the spikes estimated due to water vapor in the stratosphere using a 183 GHz line is of the order of a few hundred degrees Kelvin in emission mode and about few thousand degrees Kelvin in absorption mode. High altitude microwave radiometric studies (Wang et al. 1983) at frequencies near 94 GHz and 183 were used to retrieve atmospheric water vapor. However, the absorption coefficient at 183.31 GHz according to Croom (1965) is presented in Section 5.10.1.

5.7 Water Vapor and Microwave Attenuation

Water vapor is perhaps one of the most important lower atmospheric constituents that often control major process in the atmosphere including the weather phenomena. Studies of the height distribution of water vapor made in various countries indicate an exponential decrease of water vapor density with height starting from the surface of the earth. Theoretical calculation of microwave attenuation coefficients for such an exponential distribution in a standard model atmosphere indicates that the contribution of water vapor content in the troposphere above 10 km is less than 1% of the total value (Evans and Hagfors 1968).

The water vapor molecule possesses a large electric dipole moment and being an asymmetric-top molecule, exhibits a complex absorption spectrum of rotational transitions with a weak line at 22.235 GHz and much stronger lines at 183 GHz and 325 GHz.

The theory of molecular absorption has been considered by Van Vleck and Weisskopf (1954), Gross (1965), and Rosenkranz (1975). Extensive comparisons by Liebe (1985) between the results of calculation and measurement have shown that in general, the attenuation due to oxygen is well described by the theory. However, for water vapor, there is a discrepancy between experiment and theory that may be accounted for by the inclusion of an empirical, non-resonant correction that depends on the square of the water vapor density. The reasons behind this have been sought in terms of hydrogen bonding of water molecules to form dimmers (Llewellyn-Jones et al. 1978) or of the water molecules clustering together (Carlon and Harden 1980); or in terms of errors in the line shape used in the calculations (Westwater and Decker 1986; Thomas and Nordstrom 1982). None of these, however, is yet able to adequately explain the observed attenuation by water vapor, and the empirical correction is still necessary to align the experiment with theory.

The calculations of zenith attenuation were made by Gibbins (1986) based on the microwave propagation model of Liebe (1985) with a Van Vleck and Weisskopf (1954) line shape. However, integrated water vapor (water vapor content in a cylinder of one square meter base and 10 km height expressed in g/m³) not only control the attenuation and group delay of microwaves and millimeter waves in satellite and line-of-sight links but also introduces background radio noises in the links, thus affecting the signal-to-noise ratio.

5.7.1 Modeling of Meteorological Parameters and Attenuation Coefficients

The water vapor density (g/m³) can be derived by using

$$\rho = \frac{e \times 1800}{8.31 T_D}$$

where e is the water vapor pressure in hecto pascals and is given by

$$e = 6.105 \left[25.22 \left(1 - \frac{273}{T_D} \right) \right] - 5.31 \log_e \left(\frac{T_D}{273} \right) \tag{5.29}$$

The absorption (cm^{-1}) of water vapor molecule at 22.235 GHz is given by Equations (5.27) and (5.26).

The height distribution of the meteorological parameters such as atmospheric pressure, temperature, water vapor pressure, and water vapor densities can be fitted with the following empirical relations:

$$e(mb) = e_0 \exp(-m_1 h)$$

$$\rho \left(\frac{g}{m^3} \right) = \rho_0 \exp(-m_2 h) \tag{5.30}$$

$$P(mb) = P_0 \exp(-m_3 h)$$

$$T(K) = T_0 m_4 h$$

Besides these, the height distribution of the absorption coefficient (dB/km) can also be fitted as

$$\alpha = \alpha_0 (-m_5 h)$$

The scale factors of this fitted equation for different seasons in Kolkata (22° N) are presented in Table 5.1. The monthly variation of attenuation at 22.235 GHz in dB and the corresponding surface water vapor density over Kolkata are shown in Figure 5.6.

It is found that the nature of the monthly variations of these two parameters is the same and shows a maximum in the months of July through August. The corresponding antenna temperature may be calculated by using the relation (Allnut 1976)

$$A(dB) = 10 \log_{10} \frac{T_m - T_{cosmic}}{T_m - T_a(f)} \tag{5.31}$$

where $T_a(f)$ represents the antenna temperature at frequency f, and T_m is the mean atmospheric temperature and is considered to be 275 K. It should be emphasized that T_m is found to be dependent on frequency and ground temperature (Mitra et al. 2000).

Now, remembering Equation (5.30), we rewrite water vapor content as

$$W = H_\rho \times \rho_0 \times 10^3 \left(\frac{g}{m^2} \right) \tag{5.32}$$

TABLE 5.1

Scale Factor of the Fitted Meteorological Parameters

Parameter	Season	Surface	Scale Factors
Vapor pressure (mbar)	Pre-monsoon	28.60	0.518
	Monsoon	39.90	0.473
	Post-monsoon	16.94	0.479
	Winter	16.94	0.479
Vapor density (g/m3)	Pre-monsoon	20.96	0.492
	Monsoon	28.77	0.450
	Post-monsoon	16.78	0.410
	Winter	12.76	0.470
Pressure (hpa)	Pre-monsoon	1011.3	0.120
	Monsoon	1014.8	0.124
	Post-monsoon	1009.85	0.116
	Winter	1014.5	0.119
Temperature (K)	Pre-monsoon	302.91	6.46
	Monsoon	302.50	5.68
	Post-monsoon	297.64	4.79
	Winter	294.10	5.13
Attenuation coefficient (db/km)	Pre-monsoon	0.870	0.389
	Monsoon	1.173	0.342
	Post-monsoon	0.702	0.311
	Winter	0.540	0.374

Now with a view to explore a relationship between the integrated water vapor content and the antenna temperature measured by the radiometer, a scatter plot between the two parameters has been drawn as shown in Figure 5.7.

A regression analysis (Karmakar et al. 1999) has also been made and the best linear equation for Kolkata was found to be

$$W\left(\frac{g}{m^2}\right) = 612T_a + 16400$$

where T_a is the antenna temperature in degree Kelvin. In this connection, it may be mentioned that the same regression analysis was done by Bhattacharya (1985) for the Northern station at New Delhi and the best fit line for Delhi, India was

$$W\left(\frac{g}{m^2}\right) = 588.23T_a - 2110$$

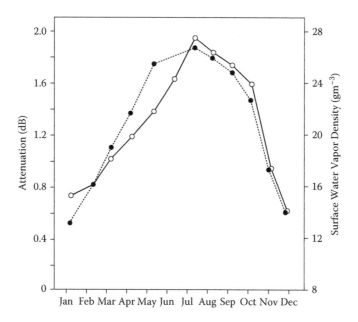

FIGURE 5.6
Monthly variation of calculated attenuation (dB) and the corresponding surface water vapor density (gm⁻³).

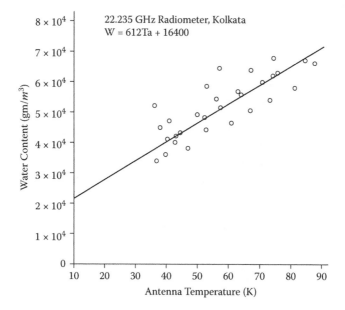

FIGURE 5.7
A scatter plot between antenna temperature and water vapor content over Kolkata.

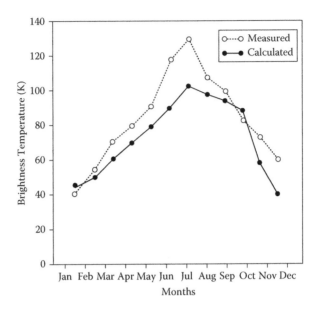

FIGURE 5.8
Monthly variation of estimated and measured brightness temperature (K) for clear air condition.

The monthly variation of brightness temperature is presented in Figure 5.8. These also bear a maximum in the months of July through August.

However, according to Raina (1981), it was found during January–March over Delhi, there is not much increase in water vapor. It starts rising from about the third week of May and continues to rise to the end of August, with broad maximum during June–August. Again, over Jodhpur (desert area), it was found that the content of water vapor becomes maximum in August and falls off around the second week of September. But, on the other hand, water vapor content over Srinagar (hilly area) increases from the first week of February and attains maximum around July–August and then falls off from the first week of September.

Now by adopting the statistical regression analyses, it was found that the calculated values of attenuation against surface vapor density were fitted well by a linear relation

$$A(dB) = -0.01106 + 0.0686\rho_s \tag{5.33}$$

with a correlation coefficient of 0.096 and rms error of 0.116 (dB). However, the results of fitting to a second-order polynomial show correlation coefficient and rms error of 0.96 and 0.116, respectively, and the relation is

$$A(dB) = -0.035 + 0.0529\rho_s + 3.8505 \times \rho_s^2 \tag{5.34}$$

From Equations (5.33) and (5.34), it is found that both linear regression and second-order polynomial regression give the same correlation coefficients and rms error, and hence linear regression may be accepted for the sake of mathematical simplicity.

The actual height distribution of water density over Kolkata (22° N, 88° E) is shown in Figure 5.9a, bearing a maximum value of about 25 g/m³ during the month of July. However, the radiosonde launched by the National Weather Service (NWS), USA, assigned to the NASA Wallops Island facility, during August 14 to 25, 1975, reveals that maximum water vapor density is 11 g/m³ (Moran and Rosen 1981) over Haystack Observatory, Westford, Massachusetts, USA. The data were then compared to find the accuracy by deploying a 22 GHz radiometer. The total uncertainty in temperature measurement was about 1.5 K in which the calibration error was 0.5 K. This is presented in Figure 5.9b.

It is to be mentioned here that while calculating the attenuation values in the microwave band, under prevailing meteorological conditions, several models

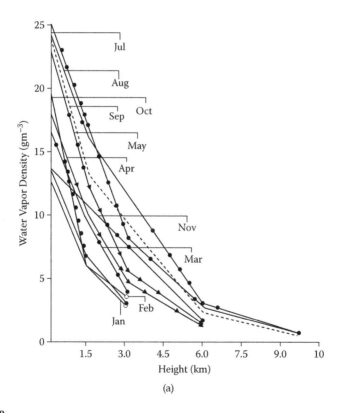

(a)

FIGURE 5.9
(a) Monthly mean profiles of water vapor density over Kolkata using radiosonde data. (b) The mean profile of the water vapor from the data of 45 radiosonde launches at Haystack Observatory in August 1975. The dotted lines denote plus and minus standard deviation.

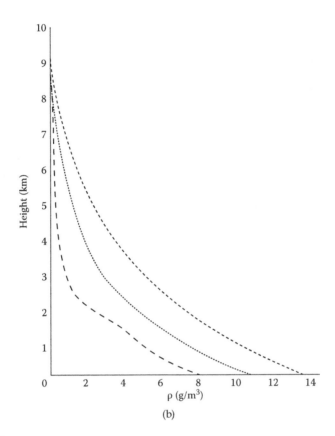

FIGURE 5.9
(Continued)

currently available in literatures may be used. Table 5.2 (Chattopadhyay 1996) is based on the model prescribed by Liebe (1989), Waters (1976), and Bhattacharya (1985). No appreciable difference among the results obtained by using different models is noticed. However, the model prescribed by Liebe is preferred for its good accuracy owing to the fact that Liebe has considered all the far-wing contributions to any desired frequency.

5.7.2 Significant Heights for Water Vapor

The well-known product-moment statistical formula may be employed to find the correlation of the temporal variation of the water vapor density at different heights in the range 0 to 10 km, averaged over 12 and 24 hours, with the variation of the integrated water vapor content. The results obtained over Digha (a coastal region) are shown in Figure 5.10. It is revealed that the correlation coefficient, between the integrated water vapor (integrated between 0 and 10 km) and the water vapor density at different heights, is 0.65 and, at

TABLE 5.2

Comparative Studies of Attenuation Using Different Models

Month	Using Liebe Model (dB)	Using Waters Model (dB)	Using Bhattacharya Model (dB)
January	0.65339	0.665	0.6143
February	0.71	0.689	0.646
March	1.065	1.072	1.035
April	1.1869	1.193	1.164
May	1.391	1.390	1.386
June	1.724	1.703	1.716
July	2.036	2.027	2.0235
August	1.927	1.880	1.915
September	1.7084	1.676	1.683
October	1.4316	1.423	1.40
November	0.9403	0.9533	0.905
December	0.5899	0.610	0.5656

surface, it attains a broad maximum of about 0.95 at the heights (2 ± 0.5) km and drops to as low as about 0.4 for a height of 10 km for the averaging time of 12 and 24 hours. For the 6-hour periods of averaging, however, there is no noticeable correlation. Thus, the heights over a range of 0.5 km centered on 2 km are found to be the most significant heights representing the diurnal variation pattern of the integrated water vapor content (Sen et al. 1989).

A comparative study of the diurnal pattern of the integrated water vapor content and water vapor density at 2 km height has been made for 24 hours as well as 48 hours as shown in Figure 5.11. It is evident from this figure that

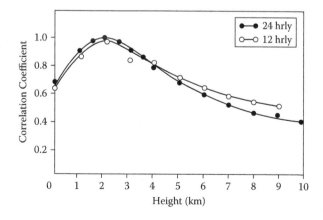

FIGURE 5.10

Correlation between the integrated water vapor content between the surface and a height of 10 km and water vapor density at several levels between the surface and a height of 10 km.

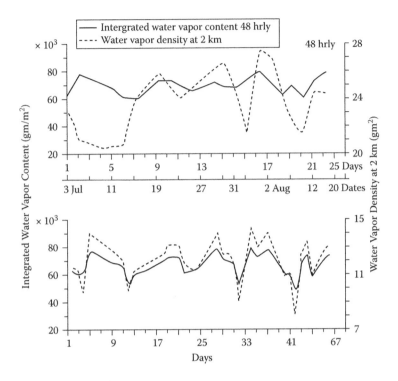

FIGURE 5.11
A comparative study of diurnal pattern of the integrated water vapor content and water vapor density at 2 km height for 24 hourly and 48 hourly averaging time.

there exists a noticeable correlation in the case of 24 hourly averaging, but 48 hourly averaging to the same produces no appreciable correlation.

The integrated water vapor content values for various heights from ground level is divided by the water vapor density found at 2 km height, which exhibits the highest correlation, as shown in Figure 5.10. The values thus obtained, known as normalized values, W, are plotted against the average values of the heights up to which the densities are integrated. This technique has been reported for the monsoon data obtained for individual sites over the Indian subcontinent. The results obtained are shown in Figure 5.12. It is interesting to note that all these curves for the monsoon period exhibit a similar variation tending to attain a constant value near the average height of 5 km, which actually corresponds to a height of 10 km up to which the integration has been carried out. However, upon furthering the regression analyses, it has been found that W_N fits with exponential model such as

$$W_N = K + ab^H$$

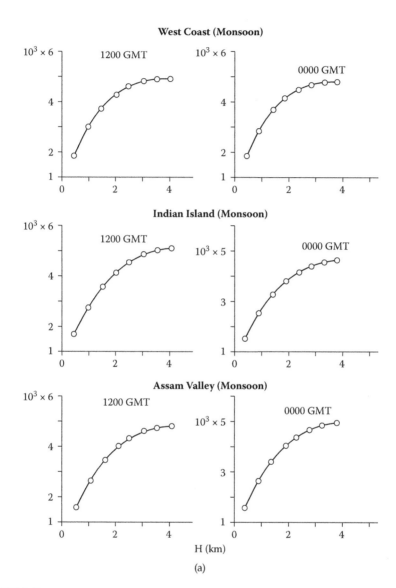

FIGURE 5.12

(a) Ratio of integrated water vapor content to water vapor density at height of 2 km plotted against the height up to which densities have been integrated on India's West Coast. (b) Ratio of integrated water vapor content to water vapor density at height of 2 km plotted against the height up to which densities have been integrated on India's Southeast Coast.

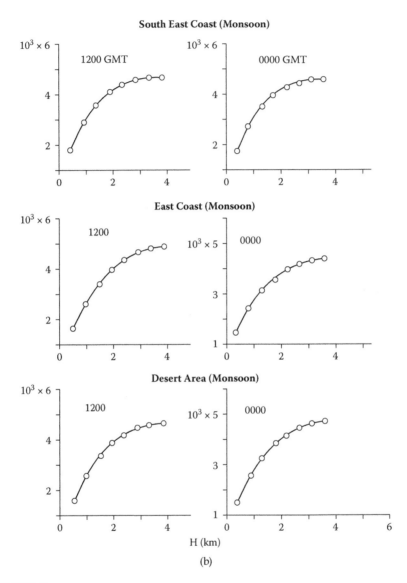

FIGURE 5.12
(Continued)

where W_N is the normalized vapor density expressed in mt at an average height H, and K, a, and b are the arbitrary constants.

5.7.2.1 Time-Scale Dependence of Vertical Distribution of Water Vapor

The observed maximum correlation between the water vapor content at 2 km height and the integrated water vapor content suggests that the integrated water vapor content is not affected by the temporal variation of water vapor density at the surface or that near 10 km height within the time scale of 12 to 24 hours. Hence, measurements around the 2 km height, which turns out to be the scale height, will be the most significant height in showing the temporal variations of the integrated water vapor content. It may be noted, however, that for a time resolution between 12 and 24 hours the correlation is good and, therefore, any transportation of the water vapor from the surface to the altitude at about 2 km must have an negligible effect within the time scale, in order that the observed lack of correlation of water vapor densities between the surface and the upper height regions may occur. If the time scale is increased to 48 hours, the integrated water vapor content is found to be poorly related with that at around 2 km height. This difference of behavior in the nature of the correlation for a short (24 hours) and a long (48 hours) time scale suggests the transportation of water vapor to higher altitudes carried within a time scale greater than 24 hours. For a shorter time scale of 6 hours, again correlation is insignificant. This suggests the presence of finer and independent components of the temporal variation at various heights.

Further, the tendency of W_N, the normalized water vapor density distribution, to attain a constant value, as shown in Figure 5.12, suggests that an increase in the altitude, up to which the water vapor densities are integrated, beyond 10 km will have a negligible effect on the integrated water vapor content.

So it appears that the measurement of water vapor density at the significant height 2 km alone may be adequate for obtaining a clear picture of the temporal variation of the integrated water vapor content in the atmosphere within the time scale of 12 to 24 hours. It is to be noted that for a longer time scale, the transportation of water vapor from the surface to higher altitudes is to be considered.

It may also be noted that the radiosonde data covers only nonrainy periods. But due to rain, for which data were not available, there may be an appreciable rise in water vapor density at cloud heights (2–5 km) as well as at the heights encompassing rain.

The water vapor density at different height in the atmosphere may also be derived by using Equations (5.28) and (5.29). It is revealed from the height distribution of water vapor density that the maximum value of 28.77 g/m3 is obtained in August and the minimum of 11.80 gm–3 in January, as observed from the fitted curves of height distribution of water vapor density over Kolkata (a tropical station), India (Karmakar et al. 1999).

Fitting of the density data with an exponential equation exp (–h/b) reveals that the monthly variation of scale height of water vapor becomes maximum during the months of July and August as is expected from the monthly variation of water vapor density over Kolkata. This is presented in Figure 5.13. Here, the water vapor scale height is defined as the height at which water vapor density becomes $1/e$ times the surface value. The average value of calculated water vapor scale height over Kolkata (Figure 5.13), by using radiosonde data, is 2.45 km.

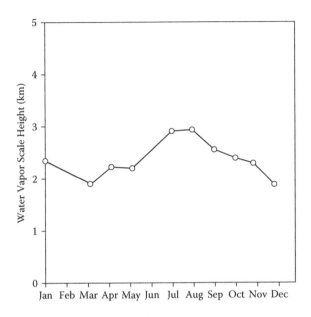

FIGURE 5.13
Monthly variation of estimated water vapor scale height over Kolkata (for clear conditions only).

However, referring back to Figure 5.8, the profile has approximately an exponential dependence with a scale height of 2.2 km, although the high altitude part decreases more slowly than an exponential. Individual profiles deviated greatly from exponential dependence.

It is to be noted that the standard atmosphere is taken into consideration; Equation (5.30) may not be the actual representation. It may so happen that most of the times during the radiosonde studies, the sky may be overcast with thin or thick clouds bearing nonprecipitable liquid water. This, in turn, suggests the use of continuous monitoring of water vapor by deploying microwave radiometers.

5.7.3 Frequency Dependence of Water Vapor Distribution

Molecular resonance absorption or emission peaks are generally exploited for remote sensing of the atmospheric gases. Westwater and Decker (1977) pointed out that the operation of a receiver at the resonance peak may lead to serious errors in the retrieval mechanism due to pressure broadening of rotational lines. This pressure-broadening effect is predominant at 22.235 GHz, where the line is relatively weaker as compared to the line at 183 GHz, and the contribution of the wings of the higher frequency lines is also significant even at the peak of the water vapor resonance line (Hogg et al. 1983). Westwater (1967) suggested some offset frequency of operation, such as 21.0 or 24.4 GHz, as the operating frequency where the change in absorption caused by pressure broadening would be insignificant. At 21.0 GHz, the pressure-broadened line is approximately two-thirds of its maximum emission intensity. Measurements at the offset frequency are, in fact, less sensitive to the distribution of water vapor with altitude but are better correlated with the integrated values of water vapor. But for the standard atmosphere, the best possible choice of frequency is found to be 23.85 GHz as at this frequency the absorption (dB/km) is not altitude dependent.

The pressure-broadening effect is caused by pressure induced molecular collisions between H_2O and N_2 molecules and further increased by the temperature of the medium. For places like Antarctica, where both the surface water vapor concentration and temperature are at lower values, the pressure broadening may not contribute significantly to affect the radiometric measurement even at the line center 22.235 GHz, and retrieval of water vapor density might be fairly accurate.

5.8 Choice of Frequency

In atmospheric remote sensing applications, the selected frequency should be very sensitive to the atmospheric parameters to be measured. Hogg et al. (1983) discussed in detail the choice of frequency and prescribed some offset frequency away from 22.235 GHz as appropriate for all locations. However, calculation of brightness temperatures (Resch 1983) around the 22 GHz water vapor line with a surface temperature of 30°C, standard lapse rate, and constant relative humidity 81.6% for $0 < h < 1000$ m and 99% for 100 m $< h < 3800$ m (Figure 5.14) shows that the high altitude profile is sharper than the low altitude profile, keeping water vapor content 2 g/m2 the same in both the cases. This suggests that if one wishes to maximize the signal from a given amount of vapor, then the observation should be carried at 22.235 GHz. It is also clear from the Figure 5.14 that a single frequency measurement of brightness temperature near the half-power point of the line profile would provide the most accurate estimates. Moreover, as the line shape function remains almost unchanged with respect to height in regions with dry climates, the use of the 22.235 GHz resonance line for radiometric measurement of water vapor density would be fairly accurate for regions like Antarctica.

Thus the choice of frequency depends on the site, season, and local meteorological condition. However, in selection of operating frequency, the following criteria must be taken into consideration (Dutta and Sen 1994):

1. The brightness temperature should be sufficiently sensitive to water vapor density and weakly sensitive to other atmospheric variables, namely, temperature, pressure, and height.

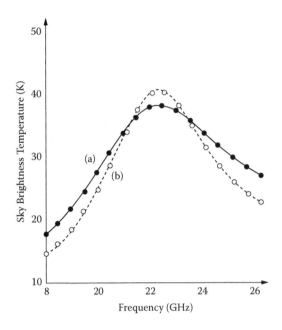

FIGURE 5.14
Line profiles of atmospheric water vapor for two different vertical distributions in standard atmosphere (a) RH = 81% for 0 < H < 1000 m (open circles), (b) RH = 99% for 1000 < H < 3800 m (filled circles).

2. For integrated water vapor measurement, the water vapor weighting function at the selected frequency should be height independent, that is, the weighting function should have a constant profile with height.

3. For vertical profiling of water vapor, the water vapor density weighting function should be sufficiently different with respect to height.

However, according to Raina and Ghosh (1993), Figure 5.15 shows zenith antenna temperatures calculated at various frequencies for surface values of the water vapor at 5, 10, 15, and 20 g/m3. It is observed that for water vapor content of 5 g/m3, antenna temperature varies between 2 K and 22 K. Likewise, variations for water vapor content at 10, 15, and 20 g/m3 can be determined from Figure 5.15. Antenna temperature, as expected, is negligible below 10 GHz. But it increases with an increase in frequency, thereafter becoming maximum at the water vapor resonance line of 22.235 GHz, and then decreases with a further increase in frequency. The variation of the antenna temperature with frequency at different values of surface water vapor content is shown in Figure 5.16. The maximum value of antenna

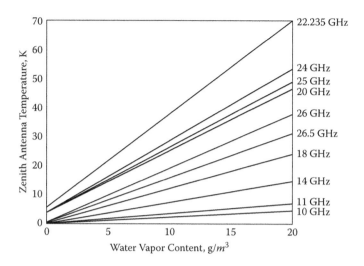

FIGURE 5.15
Zenith antenna temperature as a function of surface water vapor content.

temperature is found to be around 75 K for water vapor content of 20 g/m3 at 22.235 GHz and the minimum around 22 K at 5 g/m3.

Figure 5.17a shows the variation of water vapor density weighting function at Kolkata estimated from radiosonde values. Weighting function profiles are drawn for frequencies at 21.0 and 22.235 GHz. The 21.0 GHz

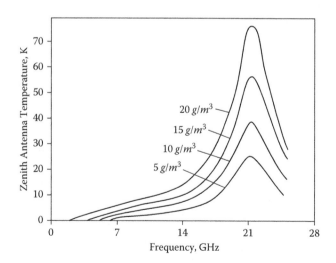

FIGURE 5.16
Variation of zenith antenna temperature with frequency for various values of water vapor content.

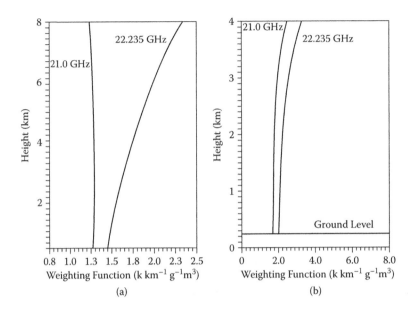

FIGURE 5.17
(a) Variation of weighting function at 21.0 and 22.235 GHz over Kolkata. (b) Variation of weighting function at 21.0 and 22.235 GHz over Antarctica.

profile indicates a more or less constant trend with height, which favors the selection of 21.0 GHz for retrieval of integrated water vapor at tropical regions. Figure 5.17b represents the estimated variation of water vapor density weighting function with height at Antarctica up to 4 km. At any selected frequency in microwave band, the maximum absorption depends on the (1) line strength factor and (2) line width parameter. Their values are decided mainly by collisions between H_2O and N_2 molecules. The effects of H_2O–H_2O collisions are very small and for Antarctica they are negligible. In Figure 5.17 the water vapor density weighting functions as estimated for Antarctica, at 21.0 and 22.235 GHz, are nearly constant through the first 3 km of height. It has been found that 70%–80% of the attenuation is caused by the lower 3 km of atmosphere. Beyond 3 km, the weighting function tends to vary slightly for both 21.0 and 22.235 GHz lines. The cause of bending may be due to the reason that radiosonde has poor accuracy in the relative humidity measurement, especially at high altitude with low humidity. Apparently, both frequencies are suitable, as they exhibit similar trends, but 22.235 GHz may be preferred for its higher sensitivity to water vapor. Therefore, a 22.235 GHz radiometer can provide reasonably good information on integrated water vapor in regions with dry climates. Measurements at the line center frequency of 22.235 GHz have in fact been reported by Decker et al. (1978).

5.9 Attenuation Studies in 50–70 GHz Band

The 50–70 GHz band centered on the 60 GHz oxygen absorption band has drawn the attention of many investigators apparently due to high absorption in the band allowing reliable communication. Near the ground level, the collision or pressure broadening (self as well as foreign gas broadening) results to merge the large number of lines (nearly 44) near 60 GHz into a single broad absorption band. In fact, at high altitudes most of the transitions are clearly resolved in this millimeter wave band. In the frequency range 56–64 GHz, there exists the level of attenuation, which is in excess of 10 dB/km at a pressure of 1013 mb, prevailing near the ground level. On either side of the strong oxygen absorption band 56–64 GHz (where attenuation rate is >10 dB/km), two bands may be considered, among which 50–55 GHz band is on the lower wing and 65–70 GHz is on the higher wing of strong oxygen absorption band where the attenuation is much less but depends markedly on the choice of exact frequency. Of these two bands the 50–55 GHz has been more widely studied as the instrumentation is more cost effective at this band compared to that at 65–70 GHz where the rain attenuation is also much higher (Karmakar et al. 1994).

However, the influence of water vapor is small in the so-called oxygen band because of large frequency separation between vapor and oxygen absorption. It is also known that the effect of clouds depends on the liquid water content and underlying surface emissivity, which is analogous to vapor effect. Clouds can have appreciable water content and therefore can change the weighting function and hence the brightness temperature also. According to Guiraud et al. (1979), content of liquid water is found to decrease below the freezing point as a result of which clouds mainly affect the lowest sensing channels in the polar regions; but on the other hand, clouds in the mid latitudes and tropics can affect the channels with higher peaking weighting functions (Gibbins 1988). But clouds over the ocean can either increase or decrease the brightness temperature with no effect at 53.7 GHz. Over high emissivity lands, the clouds always decrease the brightness temperature. It is to be mentioned here that the study of the 50–70 GHz band is made useful for temperature sensing, but the effect of water vapor and liquid water in the absorption band is to be studied rigorously over the place in question. A sample plot of calculated values from NOAA 1981–1982 data is shown in Figure 5.18 which shows that Kolkata and Srinagar being the two extreme locations, with widely varied local weather patterns, are having nearly the same specific attention profile within the frequency band of 50–55 GHz in clear weather conditions. Even for Kolkata in January and August, which exhibit extreme integrated water vapor contents in the atmosphere, 25.1 kg/m2 and 66.50/m2 respectively, and corresponding to surface vapor contents of 11.2 gm/m3 and 24.5 gm/m3, respectively, the zenith opacity or the total integrated attenuation and specific attenuation varies

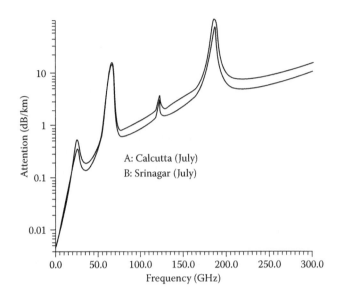

FIGURE 5.18
A sample plot of calculated specific attenuation versus frequency of propagation at Kolkata (Lat 22.34° N) and Srinagar (Lat 34.06° N).

respectively only within 0.5 dB and 0.2 dB/km for any frequency between 50 to 55 GHz (Figure 5.19).

In this context, it is worthwhile to mention that during a theoretical study made over the Indian subcontinent to identify the window frequency band, the 50–55 GHz band was found to be invariant, as its specific attenuation remains more or less invariant irrespective of locations. A sample plot is shown in Figure 5.20 for clarity. The propagation parameters in 50–70 GHz in clear air and in the presence of hydrometeors, the distinct transition starts to occur at about 300 mb pressure in the range 57–63 GHz band. The sharp transition in this band is useful for communication. This seems to be plausible because 58 GHz attenuation is 4 times larger than that at 59 GHz attenuation. India being a tropical country, most of its regions especially the coastal regions and to some extent the state adjoining these regions, remains very much humid for a substantial period of a year. In the lower troposphere, the gaseous absorption, especially oxygen and water vapor dominates in the 30–300 GHz band. But as we know, the density of oxygen is constant unlike water vapor in the lower atmosphere; the oxygen absorption peak at 60 GHz would show consistency in its propagation behavior unlike of that around water vapor absorption peaks. Thus in the oxygen band as absorption coefficient is strongly dependent on the partial pressure of oxygen in the atmosphere, and has a uniform and time-invariant mixing

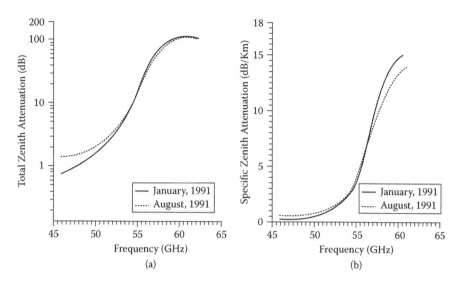

FIGURE 5.19
The variation of (a) total zenith attenuation (b) specific attenuation in clear weather condition within 50–55 GHz band as calculated from data available for Kokata, India.

ratio, underlies the choice of oxygen band for retrieving purposes (Ulaby et al. 1981). According to Westwater and Decker (1977), the vertical profile of weighting function of 55.45 GHz decreases more rapidly with increasing height than the lower frequency (52.85, 53.85 GHz) profiles, which implies that the surface layers within 1 km has the greater influence on 55.45 GHz

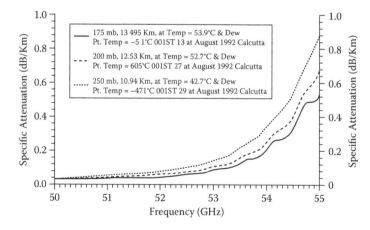

FIGURE 5.20
The plot gives the nature of the variation for the specific attenuation values at different frequencies between 50 to 55 GHz for three discrete radiosonde data sets (as indicated through inset) for Kolkata, which shows the existence of different O_2 absorption lines in the 60 GHz band.

and the least influence on 52.85 GHz. Also at 55.45 GHz, it is almost dominated by the first 3 to 4 km of the lower atmosphere. The magnitude of the weighting function at 52.85 GHz remains significant up to 10 km height and even higher.

The calculation made by Karmakar et al. (1994), regarding clear air attenuation based on analytical equations developed by Ulaby et al. (1981), in the frequency band 50–55 GHz for its propagation through earth's atmosphere is presented in Figure 5.20. A more detailed description for millimeter wave absorption in the atmosphere can be achieved by using the MPM model (Liebe 1989).

The absorption thus produced in dB/km (Figure 5.21) at P = 1013 mb shows that the highest level of attenuation occurs at 55 GHz, which is approximately 10 times larger than that at 50 GHz. In this case, it is to be noted that during calculations, atmospheric pressure (P = 1013 mb) has not been altered but the temperatures range from 1.9°C to 34.6°C as is recorded over the whole of the country. Here, 19°C is found to be the minimum and 34.6°C is the maximum during the whole of the year over India. The difference in attenuation is

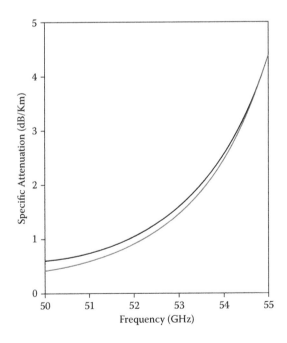

FIGURE 5.21
The plot shows the variational pattern of calculated specific attenuation between 50 and 55 GHz at two extreme climatic conditions of two locations of India. Black line indicates for Madras (lat 13° N) and grey line for Srinagar (lat 34° N). Both these specific attenuations are calculated at P = 1013 mb, T = 34.6°C.

pronounced, roughly more than 1 dB/km, at 50 GHz and is going to decrease at higher frequencies, with an intersection at 53.75 GHz. This result signifies 53.75 GHz is unique over the Indian subcontinent for pressure sounder due to the fact that attenuation remains independent of temperature around this frequency.

It is also observed that the temperature dependence of attenuation, in terms of percentage change/degree centigrade from 1.9°C to 34.6°C, is linear over the frequency range 50–55 GHz. This conforms with the results obtained by Gibbins (1988).

According to Karmakar et al. (1994), the water vapor attenuation and oxygen attenuation were calculated separately and then summed to find out total attenuation coefficient for the frequency range 50–55 GHz. It is found there that the water vapor attenuation over Chennai, India, ranges from 0.316 to 0.367 dB/km and corresponding oxygen attenuation ranges from 0.266 to 3.765 dB/km for temperature $t = 22.0$°C. But on the other hand, water vapor attenuation over Srinagar ranges from 0.077 to 0.09 dB/km and oxygen attenuation from 0.388 to 4.324 dB/km for temperature $t = 1.9$°C, for the frequency range 50–55 GHz. It is clearly understood from the results that the lower the dew point temperature, the larger is the dominance of oxygen attenuation over the total attenuation. For example, at 55 GHz the ratio of water vapor and oxygen attenuation over Chennai and Srinagar comes out to be 0.097 and 0.020, respectively, which clearly signifies the dependence of dew-point temperature on total attenuation coefficient. Also it is seen that attenuation varies from 0.45 to 4.5 dB/km for 50 GHz and 55 GHz, respectively.

5.9.1 Possible Application in 50–70 GHz Band

From the discussions made in the previous sections, it appears that the oxygen band may be precluded from usage with communications. But this large attenuation provided by the atmosphere gives additional isolation, which may be used by the communication engineers for mobile networking in rural or urban areas. This oxygen band may also be used for traffic control and vehicular collision avoidance radar. It has been well discussed that the distinct transition starts to occur in frequency band 57–63 GHz at about 300 mb pressure. In this band attenuation ranges from 1 dB to 100 dB at the central frequency, that is, around 60 GHz. Apparently this large attenuation does not favor earth–space communication.

It is also observed by Karmakar et al. (1994) that the effect of variation of water vapor content in the atmosphere does not have much effect on attenuation due to gaseous molecules in the 50–55 GHz band at a particular location over the Indian subcontinent. This primarily could be the major reason for exploiting this band for communication purposes over the Indian subcontinent.

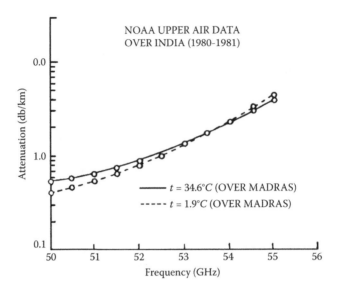

FIGURE 5.22
Clear air attenuation coefficient in the frequency band 50–55 GHz. Curves were drawn at surface pressure 1013 mb. Intersection of the curves occur at 53.75 GHz. This is considered as a frequency to be selected for pressure sounding over the Indian subcontinent.

5.10 Attenuation Studies at 94 GHz

In the neighborhood of 94 GHz, the absorption due to water vapor is not only contributed by two distinct rotational lines at 22.235 and 183.31 GHz together, but also contributed by the nonresonant absorption process. However, the major contribution of water vapor attenuation at 94 GHz is due to rotational absorption lines of 22.235 and 183.31 GHz. Thus, the attenuation due to water vapor at 94 GHz may be estimated by summing the contributions of the two aforementioned absorption lines. The contribution of attenuation coefficients at the upper wing of 22.235 GHz and those at the lower wing of 183.31 GHz at different heights were summed up, which, in other words, provide the attenuation rate at 94 GHz (Sen et al. 1988). The contributions of higher frequency resonant lines have been neglected.

5.10.1 Theoretical Consideration

The attenuation rate α (cm–1) due to water vapor in the atmosphere at 22.235 GHz can be calculated on the basis of Equations (5.24) through (5.28). Further, the expression for attenuation rate at 183.31 GHz water vapor line $\alpha_{(1.64\ mm)}$ using the appropriate term values (Denison 1940) and temperature exponent

(Benedict and Kaplan 1954) is shown by Croom (1965).

$$\alpha_{183.31} = \frac{0.646 N^2 v \exp(-200/T)}{10^{28} T^{2.5}} \left[\frac{\Delta v_{P(183)}}{\{(v-v_0)^2\} + \Delta v_{P(183)}^2 \{(v-v_0)^2\} - \Delta v_{P(183)}^2} + \frac{\Delta v_{P(183)}}{} \right]$$

$$+ \frac{1.8}{10^{52}} \frac{N v^2}{T^{1.5}} \Delta v_{P(183)} \tag{5.35}$$

where v_0 (GHz) is the resonance frequency for the 183.31 GHz line, v and T bear the same significance as mentioned earlier, and N represents the number of water vapor molecules per cubic centimeter and is given by

$$N = \frac{\rho}{18} \times 6.023 \times 10^{17} \tag{5.36}$$

Here ρ is the water vapor density in g/m3.

The line half width parameter for 183.311 GHz, $\Delta v_{P(183)}$, as given in Waters (1976), can be represented as

$$\Delta v_{P(183)} = 2.68 \left\{ \frac{P}{1013} \right\} \left\{ \frac{300}{T} \right\}^{0.649} \{1 + 0.0203 \rho T / P\} \tag{5.37}$$

At 94 GHz, the water vapor attenuation could be assumed primarily due to the contribution of resonant and nonresonant parts of 22.235 and 183.31 GHz water vapor absorption spectra as given in Equations (5.24) and (5.35), respectively, and hence is calculated accordingly. It may be mentioned here that the contribution from the lower wing of the water vapor line at 325.153 GHz is found to be much less compared to the attenuation value at 94 GHz. The attenuation contributions at 94 GHz due to 325.153 GHz as well as due to other high resonant absorption lines in near infrared have not, therefore, been taken into account.

5.10.2 Computation of 94 GHz Attenuation from Radiosonde Data

The radiosonde data for the months of July and August 1979 during the MONEX (Monsoon Experiment) period over Digha (lat 21°41′ N, long 87°40′ E), a coastal region near the Bay of Bengal, and that for the same period for Kolkata, four times a day at 00.00, 06.00, 12.00, and 18.00 GMT were used to calculate N, ρ, e, and $\Delta v_{P(22)}$ and $\Delta v_{P(183)}$ finally to obtain the value of α at 94 GHz as a function of altitude. Tables 5.3 and 5.4 show sample tabulations for all the parameters for a particular day and time for two different years over Kolkata and Digha. The variation of line width

TABLE 5.3

Radio Meteorological Data for Kolkata (22° N) at 00 GMT on July 19, 1991, along with Water Vapor Absorptions at 22.235 and 183 GHz Parameters

Height (km)	Pressure (mb)	Temperature (°C)	Dew Point (°C)	Vapor Density (g/m^3)	$\Delta\nu_{22}$ (GHz)	$\Delta\nu_{183}$ (GHz)	Number of Molecules (cc)	$(\alpha_{94})_{22}$ (dB/km)	$(\alpha_{94})_{183}$ (dB/km)	$(\alpha_{94})_{22} + (\alpha_{94})_{183}$	$(\alpha_{94})_{183}/(\alpha_{94})_{22}$
0	1000	29.8	26.1	24.81	3.164	3.030	8.304E+17	0.62	0.87	1.49	1.4
0.465	950	27.8	27.6	27.00	3.067	2.944	9.036E+17	0.66	0.93	1.59	1.4
0.945	900	24.6	22.6	20.31	2.843	2.179	6.795E+17	0.47	0.66	1.13	1.4
1.446	850	22.6	21.1	18.60	2.686	2.568	6.226 E+17	0.41	0.57	0.98	1.4
1.974	800	19.0	18.6	16.05	2.523	2.409	5.370 E+17	0.34	0.47	0.81	1.4
2.528	750	15.4	15.4	13.23	2.355	2.245	4.427 E+17	0.26	0.37	0.64	1.4
3.112	700	12.5	11.9	10.66	2.185	2.080	3.566 E+17	0.20	0.28	0.48	1.4
3.734	650	10.0	8.0	8.32	2.016	1.917	2.784 E+17	0.15	0.20	0.35	1.4
4.397	600	7.1	4.6	6.67	1.857	1.746	2.231 E+17	0.11	0.15	0.26	1.4
5.110	550	3.6	1.2	5.31	1.704	1.616	1.777E+17	8.17 E-02	0.11	0.20	1.4
5.880	500	-0.2	-2.5	4.12	1.551	1.417	1.379 E+17	5.89 E-02	8.28 E-02	0.14	1.4
6.718	450	-5.1	-6.3	3.15	1.403	1.329	1.054 E+17	4.18 E-02	5.89 E-02	0.10	1.4
7.635	400	-9.9	-12.1	2.06	1.250	1.184	6.892 E+16	2.50 E-02	3.52 E-02	6.02E-02	1.4
8.653	350	-15.9	-19.3	1.18	1.101	1.041	3.955 E+16	1.30 E-02	1.84 E-02	3.15 E-02	1.4
9.794	300	-25.5	-29.9	0.55	0.958	0.906	1.642 E+16	4.97 E-03	7.07 E-03	1.20 E-02	1.4

TABLE 5.4

Radio Meteorological Data for Digha (21° N) at 00 GMT on July 19, 1991, along with Water Vapor Absorptions at 22.235 and 183 GHz Parameters

Height (km)	Pressure (mb)	Temperature (°C)	Dew Point (°C)	Vapor Density (g/m³)	$\Delta\nu_{22}$ (GHz)	$\Delta\nu_{183}$ (GHz)	Number of Molecules (cc)	$(\alpha_{94})_{22}$	$(\alpha_{94})_{183}$	$(\alpha_{94})_{22} + (\alpha_{94})_{183}$	$(\alpha_{94})_{183}/(\alpha_{94})_{22}$
0.471	950	26.3	24.0	22.03	3.001	2.871	7.371E+17	0.53	0.75	1.28	1.4
0.947	900	23.6	19.2	16.63	2.794	2.665	5.565E+07	0.38	0.53	0.91	1.4
1.443	850	20.7	16.4	14.06	2.629	2.504	4.705E+17	0.31	0.43	0.73	1.4
1.996	800	17.8	13.4	11.70	2.465	2.345	3.915E+17	0.24	0.34	0.58	1.4
2.512	750	14.7	9.6	9.22	2.30	2.18	3.085E+17	0.18	0.25	0.43	1.4
3.1	700	11.0	5.7	7.17	2.14	2.03	2.398E+17	0.13	0.19	0.32	1.4
3.710	650	7.9	2.3	5.73	1.99	1.88	1.918E+17	0.10	0.14	0.24	1.4
4.364	600	4.2	−1.3	4.48	1.84	1.74	1.498E+17	7.40E−02	0.10	0.18	1.4
5.068	550	0.1	−5.5	3.34	1.69	1.60	1.116E+17	5.17E−02	7.27E−02	0.12	1.4
5.826	500	−3.6	−10.1	2.39	1.54	1.45	7.998E+16	3.45E−02	4.85E−02	8.29E−02	1.4
6.65	450	−8.4	−15.8	1.55	1.39	1.31	5.182E+16	2.07E−02	2.92E−02	4.99E−02	1.4

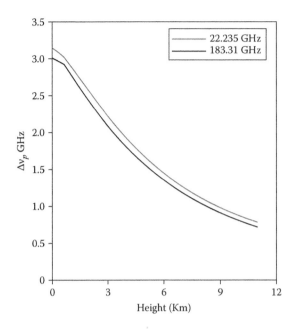

FIGURE 5.23
Plot shows the variation of line width parameters of O_2 molecule at the absorption line frequencies 22.235 and 183.31 GHz.

parameters of 22.235 and 183.31 GHz, respectively, $\Delta v_{P(22)}$, and, $\Delta v_{P(183)}$ with height is shown in Figure 5.23.

Computational techniques have been employed to find the different parameters at different heights four times a day. The following striking features are revealed from the computation:

1. The 94 GHz specific attenuation values as obtained from 22.235 GHz and 183.31 GHz absorption lines are of nearly same order of magnitude for a set of meteorological data at a particular height.

2. The 94 GHz attenuation value as obtained from 183.31 GHz is on average 1.4 times that obtained from 22.235 GHz line (refer to last column of Tables 5.3 and 5.4). Hence the following empirical relation can be formulated:

$$\alpha_{94} = (\alpha_{94})_{22} + (\alpha_{94})_{183}$$

where α_{94} is the total water vapor attenuation due the contributions from 22.235 and 183.31 GHz collectively. $(\alpha_{94})_{22}$ and $(\alpha_{94})_{183}$ are the nonresonant water vapor attenuation contributions at 94 GHz.

Now referring to Table 5.3, it clearly indicates that

$$(\alpha_{94})_{183} = 1.4 \times (\alpha_{94})_{22}$$

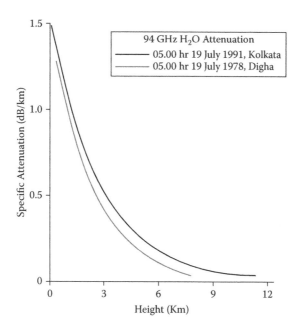

FIGURE 5.24
The plot shows the specific attenuation (α) due to water vapor at 94 GHz as calculated from 22.235 and 183.31 GHz water vapor absorption lines (from radiosonde data of July 19, 1991, for Kolkata, and July 19, 1978, [MONEX] for Digha [long 87°40 E, lat 21°41′ N]).

Therefore,

$$\alpha_{94} = (\alpha_{94})_{22} + 1.4 \times (\alpha_{94})_{22} = 2.4(\alpha_{94})_{22} \tag{5.38}$$

This reveals an interesting result: water vapor attenuation at 94 GHz is 2.4 times that contributed from the line at 22.235 GHz alone. By using Equation (5.38) a plot of 94 GHz attenuation at respective heights is represented in Figure 5.24. The plot assumes a best fit negative exponential decay curve, which is represented as

$$\alpha = C_0 exp\left\{-h\big/C_1\right\} \tag{5.39}$$

where C_0 and C_1 are regression constants. The scale height C_1 is found to be 1.93 and 2.68 km for Digha and Calcutta, respectively. This again shows that even the use of the same data set of the same day for two different days for two different years 1978 and 1991 give different scale heights for 94 GHz specific attenuation because water vapor concentrations are different for two locations. All these results are obtained during the most humid monsoon months over the areas where water vapor density attains almost the highest values over the Indian subcontinent.

Direct radiometric measurements, however, is expected to yield higher values of water vapor attenuation due to some anomalous absorption at millimeter wave lengths. According to Liebe et al (1973) and Raydov and Furashov (1972), the water vapor attenuation in reality depends quadratically on water vapor density and temperature, which may produce higher values than those obtained by considering monomer model as discussed here. Experimental evidences show that up to 220 GHz, the measured values may exceed the calculated values by 30 to 40 percent in dB.

References

Allnut, J. E., 1976, Slant path attenuation and space diversity results using 11.6 GHz radiometer, *Proceedings of IEEE*, 123, 1197–1200.

Altshuler, E., Lamers, H. W., 1968, A troposcatter propagation experiment at 15.7 GHz over 500 m path, *Proceedings of IEEE*, 56, 1729–1731.

Anderson, W. L., Beyers, M. J., Fannin, B. M., 1959, *TRANS, IRE*, PGAP, AP-7, 258–260.

Artman, J. O., 1953, *Absorption of microwave by oxygen in millimeter wave length region*, Columbia Radiation Laboratory Report.

Barrett, A. H., Chung, V. K., 1962, A method for determination of high altitude water vapor abundances from ground based micro wave observations, *Journal of Geophysical Research*, 61, 11, 4259–4266.

Becker, E. R., Autler, S. H., 1946, Water vapor absorption of electromagnetic radiations in the cm wave length range, *Physics Review*, 70, 300–307.

Benedict, W. S., Kaplan, L. D., 1959, Calculation of line width in $H_2O–N_2$ collisions, *Journal of Chemical Physics*, 30, 385–388.

Bhattacharya, C. K., 1985, *Microwave radiometric studies of atmospheric water vapor and attenuation measurements at microwave frequencies* (PhD thesis), Benaras Hindu University, India.

Carlon, H. R., Harden, C. S., 1980, Mass spectroscopy of ion induced water clusters: An explanation of the infrared continuum absorption, *Applied Optics*, 19, 1776–1786.

Carter, C. J., Melehell, R. L., Reber, E. E., 1968, Oxygen absorption measurements in the lower atmosphere, *Journal of Geophysical Research*, 73, 3113–3120.

Chattopadhyay, S., 1996, *Radiometric Studies of the atmosphere at millimeter wave length* (PhD thesis), Calcutta University, India.

Chung, U. K., 1962, *Microwave spectrum of the planet Venus* (MS thesis), Massachusetts Institute of Technology, Cambridge.

Croom, D. L., 1965, Stratospheric thermal emission and absorption near 22.235 GHz rotational line of water vapor, *Journal of Atmospheric and Terrestrial Physics*, 27, 217–233.

Decker, M. T., Westwater, E. R., Guiraud, F. O., 1978, Experimental evaluation of ground based microwave radiometric remote sensing of atmospheric temperature and water vapor profiles, *Journal of Applied Meteorology*, 17, 1788–1795.

Denison, D. N., 1940, The infrared spectra of polyatomic molecules, *Review of Modern Physics*, 12, 175–240.

Dutta, S. K., Sen, A.K., 1994, A fresh consideration on the choice of radiometer operating frequency in atmospheric remote sensing, *International Journal of Remote Sensing*, 15, 2176.

Evans, J. V., Hagfors, T., 1968, *Radio astronomy*, New York: McGraw Hill.

Falcone, V. J. Jr., Wulfsberg, K. N., Gitelson, S., 1971, Atmospheric emission and absorption at millimeter wave length, *Radio Science*, 6, 3, 347–355.

Gaut, N. E., 1968, *Studies of water vapor by means of passive microwave techniques*, Technical Report No. 467, M.I.T.

Ghosh, S. N., Ashoke, K., 1983, *Microwave spectra and absorption coefficients of atmospheric gases*, Report from Ghosh Professors Lab, Department of Applied Physics, Calcutta University.

Ghosh, S. N., Edwards, H. D., 1956, *Rotational frequencies and absorption coefficients of atmospheric gases*, Report No. 282, Geophysics Research, Airforce, Cambridge Research Center.

Ghosh, S. N., Ghosh, A., 1986, *Some aspects of scattering of mm wave propagation through turbulent tropospheric region*, Presented at Symposium on Microwave Communication, IEEE, Calcutta Chapter, April 8, 1986.

Ghosh, S. N., Malaviya, A., 1961, Microwave absorption in the Earth's atmosphere, *Journal of Atmospheric and Terrestrial Physics*, 21, 243–246.

Gibbins, C. J., 1986, Zenithal attenuation due to molecular oxygen and water vapor in the frequency range 3–350 GHz, 22, 11, *Electronics Letters*, 577–578.

Gibbins, C. J., 1988, The effects of the atmosphere on radio wave propagation in the 50–70 GHz frequency band, *Journal of the Institution of Electronic and Radio Engineers*, 58 (Supplement), 6, 229–240.

Gordy, W., Smith W. V., Trambarule, R. F., 1953, *Microwave spectroscopy*, New York: John Wiley & Sons.

Gross, E. P., 1965, Shape of collision broadened spectral lines, *Physical Review*, 97, 395–403.

Guiraud, F. O., Howard, D. C., Hogg, D. C., 1979, A dual channel microwave radiometer for measurement of precipitable water vapor and liquid, *IEEE Trans. Geo Sc and Remote Sensing*, 17, 129–136.

Heitler, W., 1954, The quantum theory of radiation, 3rd ed., London: Oxford University Press, 180.

Hogg, D. C., Guiraud, F. O., Snider, J. B., Decker, M. T., Westwater, E. R., 1983, A steerable dual-channel microwave radiometer for measurement of water vapor and liquid in the troposphere, *Journal of Climate and Applied Meteorology*, 22, 789–806.

Janssen, M. A., 1993, Atmospheric remote sensing by microwave radiometry, New York: John Wiley & Sons.

Karmakar, P. K., Mitra, A., Sengupta, M. K., Dutta, K. A. K., Sen, A. K., 1994, Millimeter wave propagation over the Indian sub-continent in 50-55 GHz band: An overview, *Journal of Scientific and Industrial Research (India)*, 53, 267–274.

Karmakar, P. K., Chattopadhyay, S., Sen, A. K., 1999, Estimates of water vapor absorption over Calcutta at 22.235GHz, *International Journal of Remote Sensing*, 20, 2637–2651.

King, G. W., Mainer, R. M., Cross, P. C., 1947, Expected microwave absorption coefficients of water and related molecules, *Physical Review*, 71, 7, 433–443.

Kundu, N., 1983, *Reference ozone over India*, Indo-US Workshop on Global Problem Proceedings.

Liebe, H. J., 1985, An atmospheric millimeter wave propagation model, *International Journal of Infrared and Millimeter Wave (USA)*, 10, 6, 631–650.

Liebe, H. J., 1989, An updated model for millimeter wave propagation in moist air, *Radio Science*, 20, 5, 1069–1089.

Liebe, H. J., Welch, W. M., Chandler, R., 1973, Lab Measurement of electromagnetic properties of atmospheric gases at millimeter wave lengths, Propagation of Radiowaves at frequencies above 10 GHz, Proc. *IEEE, Conf,* Pub. No. 98, (New York), 244–246.

Llewellyn-Jones, K. J., Knight, R. J., Gebbie, H. A., 1978, Absorption by water vapor at 7.1 cm^{-1} and its temperature dependence, *Nature*, 274, 876–878.

Mitra, A. P., 1977, *Minor constituents in the middle atmosphere*, Scientific note (ISRO-INCOSPAR-SN-03-77) Indian Space Research Organisation, Bangalore.

Mitra, A., Karmakar, P. K., Sen, A. K., 2000, A fresh consideration for evaluating mean atmospheric temperature, *Indian Journal of Physics*, 74b, 379–382.

Moran, J. M., Rosen, B. R., 1981, Estimation of propagation delay through troposphere and microwave radiometer data, *Radio Science*, 16, 235–244.

Morse, M. D. and Maier, J. P., 2011, Detection of nonpolar ions in $2\pi_{3/2}$ ststes by radio-astronomy via magnetic dipole transitions, *The Astrophysical Journal*, 732, 2, 103–107.

Raina, M. K., Ghosh, A. B., 1993, Water vapor distribution over India using SAMIR data on board Bhaskra II Satellite and its comparison with upward looking ground based microwave radiometric measurement at 22.235 GHz over New Delhi, *Indian Journal of Radio and Space Physics*, 22, 239–245.

Raina, M., Uppal, G., 1981, Rain attenuation measurements over New Delhi with a micro-wave radiometer at 11 GHz, *IEEE Trans. Ant. and Propagation*, 29, 6, 857–864.

Randall, H. M., Dennison, D. M., Ginsburg, N., Weber, L. R., 1937, H_2O infrared spectrum, *Physical Review*, 52, 160–165.

Raydov, U. Ya., Furashov, N. I., 1972, Investigation of radiowave absorption in the atmospheric transparent window $\lambda = 0.73$mm, *Izvestiya radiofizika*, 15, 1475–1479.

Resch, G. M., 1983, *Another look at the optimum frequencies for a water vapor radiometer*, TDA Progress Report (Oct–Dec), USA.

Rosenkranz, P. W., 1975, Shape of 5 mm oxygen band in the atmosphere, *IEEE Trans.*, AP-23, 498–506.

Sen, A. K., Karmakar, P.K., Das, T. K., Devgupta, A. K., Chakraborty, P. K., Devbarman, S., 1989, Significant heights for water vapor content in the atmosphere, *International Journal of Remote Sensing*, 10, 1119–1124.

Sen, A. K., Karmakar, P. K., Devgupta, A. K., Dasgupta, M. K., Calla, O.P.N., Rana, S. S., 1990, Radiometric studies of clear air attenuation and atmospheric water vapor at 22.235 GHz over Calcutta, *Atmospheric Environment*, 24A,7, 1909–1913.

Sen, A. K., Karmakar, P. K., Mitra, A., Devgupta, A. K., Sehra, J. S., Ghosh, S. N., 1988, A theoretical estimate of tropospheric water vapor attenuation at 94 GHz from radiosonde data, *International Journal of Remote Sensing*, 9, 7, 1259–1266.

Smith, E. K., Weintraub, S., 1953, The constant in the equation for atmospheric refractive index at radio frequencies, *Proc. IRE*, 41, 1053–1057.

Staelin, D. H., 1966, Measurement and interpolation of microwave spectrum of the terrestrial atmosphere near 1 cm. wavelength, *Journal of Geophysical Research*, 71, 12, 2875–2881.

Thomas, M. E., Nordstrom, R. J., 1982, The N_2-broadened water vapor absorption line shape and infrared continuum absorption, *Journal of Quantitative Spectroscopy and Radiative Transfer*, 28, 103–112.

Townes, C. H., Schalow, A. L., 1975, *Microwave spectroscopy*, New York: McGraw Hill.

Ulaby, F. T., Moore, R. K., Fung, A. K., 1986, *Microwave remote sensing*, Vol. 3, Chapter 13, Norwood, MA: Artech House.

Ulaby, F. T., Moore, R. K., Fung, A. K., 1981, *Microwave remote sensing*, Vol. 1, Chapter 5, London: Addison-Wesley.

Van Vleck, J. H., 1947a, The absorption of microwaves by oxygen, *Physical Review*, 71, 413–424.

Van Vleck, J. H., 1947b, The absorption of microwaves by uncondensed water vapor, *Physical Review*, 71, 425–433.

Van Vleck, J. H., Weisskopf, V. F., 1954, On the shape of collision broadened lines, *Review of Modern Physics*, 17, 227–241.

Wang, J. R., King, J. L., Wilheit, T. T., 1983, Profiling atmospheric water vapor by microwave radiometry, *Journal of Climate and Applied Meteorology*, 22, 5, 779–788.

Waters, J. E., 1976, In *Methods of Experimental Physics*, Vol. 12, part B, M. L. Meeks, ed., New York: Academic Press.

Westwater, E. R., Decker, M. T., 1986, Application of statistical inversion to ground based microwave remote sensing of temperature and water vapor profiles, In *Inversion methods in atmospheric remote sensing*, A. Deepak, ed., Academic Press, New York, 949–954.

Westwater, E. R., 1967, ESSA Tech Report IER-30-ITSA 30, National Technical Information Service, Springfield, Virginia.

6

Rain Attenuation and Its Application at Microwaves

6.1 Introduction

It is known that most of the water in air remains as water vapor rather than as liquid or solid hydrometeors. It is understood that if the ambient temperature at a certain height is 0°C or below, the hydrometeors are ice. But below this height where temperature is 0°C, ice starts melting and transfers into rain. But the fall of rain is prevented by the rising air currents. As water condenses, it forms ice crystals that are small enough to be supported by air currents. These particles are clubbed together until they become too heavy for rising currents to support. These heavy particles fall as rain. The weather radar can have a strong return echo from this height, which we call a bright band. However, rain is not uniform but can be approximated to a group of rain cells in the form of a cylinder of uniform rain, for simplicity, extending from the cloud base height to the ground. Typically, the diameter of this cylinder is $d_r = 3.3 \, R^{-0.08}$; R is the rain rate in mm hr^{-1}.

It was discussed in Chapter 1 that rain produces a significant attenuation on radio waves of frequencies beyond 10 GHz. When the propagation path is intercepted by the rain cells, attenuation is caused. A deep fade occurs when the rain cells fill a large section of Fresnel's ellipsoid along the propagation path.

Each water droplet may be considered an imperfect conductor. The incoming radio wave induces displacement current. As the dielectric constant of water is 80 times larger than that of air, the density of this displacement current is large. On the other hand, the density of the displacement current is proportional to frequency. This implies that the displacement current will be larger in the microwave or millimeter wave band. In addition to this absorption, the other important phenomenon is rain scattering. In fact, scattering depends on the particle size and the frequency concerned. The process of scattering is broadly classified into two categories: Rayleigh scattering caused by fog, cloud, drizzle or light rain of sizes less than 0.4 mm; and Mie scattering caused by moderate to heavy rain for drop sizes more than 0.4 mm

(Ramachandran and Kumar 2007). In fact, Rayleigh scattering occurs when the particle is electrically small and phase shift across the particle is small, that is, the particle is much smaller than the wavelength. The radiation pattern in this case is like that of a dipole and the amount of scattered energy is proportional to D^3/λ. But, on the other hand, if the size is of the order of the wavelength then Mie scattering occurs. In this regard, the Rayleigh criterion may be cited as

$\pi D/\lambda$; D is the diameter of the particle

$\pi n D/\lambda$; n is the refractive index

For example, take a raindrop of diameter 1 mm and $\lambda = 3$ mm (equivalent to 100 GHz) and $n = 1$, then the second criterion gives the value equal to one. So if we take a microwave below 100 GHz and a raindrop of the order 1 mm then the Rayleigh scattering will dominate the situation and corresponding radiation pattern will be like that of a dipole. Thus, the absorption loss and the scattering loss when summed produces attenuation.

Rain attenuation in the microwave band is considered to be the major concern in satellite communication. The prediction of rain attenuation generally starts from a known point rainfall rate statistics, considering the vertical and horizontal structures of rain cells (Karmakar et al. 1991). Having the knowledge of rain structures and by using the climatological parameters, one can estimate rain attenuation. However, it is to be mentioned that the rapid fluctuations of point rain fall intensity are not translated directly into rapid attenuation fluctuations on a radio link (Watson et al. 1987). If one considers the spatial averaging that occurs over the active volume of a radio link (within the first Fresnel zone), then the correct choice of rain gauge integration time for 10-30 GHz radio links should lie in the range 1–5 minutes (Boldtmann and Ruthroff 1974). More precisely, for an earth-space path of length 5-10 km operating at 11.5 GHz, an integration time of 1.5 minutes should be used. But normally, the use of 1 minute rain rate gives the best agreement with the available radio data. Since, in most of the cases, we are interested in establishing a data base for 1 minute rain fall, methods for conversion of τ_{min} rain falls to 1 minute rain falls must be found, both for cumulative distribution and annual extreme values. According to Watson (1982), the annual cumulative distributions of τ_{min} rainfalls may be scaled in the range 1–60 minutes using climatically independent scaling factor C_R, related to exceedance probability. For this purpose, data from Belgium, Czechoslovakia, Italy, the Netherland, and the United Kingdom were used to give the ratios shown in Table 6.1 for conversion from 60 minutes to 5 minutes rain fall. Data presented by Harden et al. (1977) were used to evaluate the scaling ratios for 5 minutes to 1 minute rain fall. Attenuation can also be estimated from the radiometric measurements of sky-noise temperature with certain assumptions made regarding the atmospheric temperature. The vertical extent of rain can also be estimated from the meteorological measurements of the

TABLE 6.1

Average C_R Values at Different Exceedance Probability

C_R	Exceeding Probability		
	0.1%	0.01%	0.001%
$\dfrac{R_{60min}}{R_{5min}}$	0.72	0.52	0.45
$\dfrac{R_{5min}}{R_{1min}}$	0.90	0.84	0.82

height of 0° isotherm and radar reflectivity measurement from which rain attenuation along the vertical path can also be estimated.

A number of prediction procedures have been developed for the earth–space path over the last decade, which are found to be applicable to temperate climates but found to overestimate rain attenuation in tropical climates (Karmakar et al. 2000). This type of overestimation is considered to be due to an incorrect estimate of the effective path length, which, in fact, essentially leads to an inaccurate estimation of path attenuation. This suggests that the equivalent vertical path through rain is not equal to the physical rain height and it should be decreased by the use of reduction factor. The International Telecommunication Union Radiocommunication Sector (ITU-R) has developed a model for the path length reduction coefficient for the projection of the path (Figure 6.1), with the vertical path equivalent to the height of 0° isotherms. An empirical reduction factor for the earth–space path has been proposed to derive an effective rain height from the height of 0° isotherms during rainy conditions.

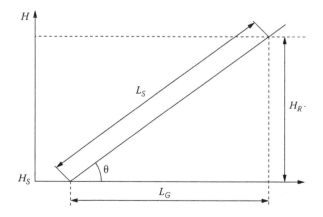

FIGURE 6.1
Schematic diagram of the earth–space path.

6.2 Radiometric Estimation of Rain Attenuation

Rain attenuation is characterized by the nonuniformity of rainfall intensity; raindrop number density, size, shape, orientation; and raindrop temperature in addition to the intrinsic variability in time and space. Rain attenuation in an earth space path may be represented by the following relation:

$$A(s) = \int_0^\infty \lambda(s)\,ds \tag{6.1}$$

where ds represent the incremental distance from the ground along the earth–space path under consideration and $\lambda(s)$ is the specific attenuation. For practical purposes, however, the calculations of rain attenuation is extended and approximated to a simple power law for a number of drop sizes distributions and temperatures in the form

$$A(S) = \int_0^{Ls} aR^b\,ds \tag{6.2}$$

where a and b are the coefficients that depend on frequency, temperature, and drop size distribution; R is the rain rate in mm hr^{-1}; and λ is identified with aR^b in dB km^{-1}. L_s is the slant path below rain height and is given by

$$L_s = \frac{H_R - H_s}{\sin\theta} \tag{6.3}$$

Here θ is the elevation angle, H_R is the rain height at the required latitude, and H_s is the height at which the radiometer is located above the sea level. Looking to Figure 6.1, we consider a new parameter, L_G, as the projection of the rainy path along the ground, which is given by

$$L_G = \frac{H_R - H_s}{\tan\theta} \tag{6.4}$$

Here it is to be remembered that the rain is assumed to be uniform along the path and that the physical rain height is assumed to be coincident with the effective rain height. Any nonuniformity in the vertical profile of rain is integrated in time, provided raindrops are falling through the ideal column reaching the ground. Exceptions may occur when drops of sufficiently comparable sizes remain aloft due to updrafts and when water is removed from the radio path by the horizontal component of wind.

The values of the coefficients a and b for 22.234 and 31.4 GHz in Equation (6.2) may be calculated using the Mie theory assuming spherical raindrops

TABLE 6.2

Values of the Constants a and b

Distribution	Frequency			
	22.235 GHz		31.4 GHz	
	a	b	a	b
LP	0.08	1.06	0.17	1.0
MP	0.09	1.087	0.209	1.033
JD (drizzle)	0.068	1.026	0.142	1.047
JT	0.132	0.918	0.301	0.812

with different drop size distributions, as discussed in Chapter 1. The Gunn and Kinzer (1949) terminal velocity of raindrops and Ray's index of refraction may be assumed with rain cell temperature of 10°C as discussed and presented in Section 1.8.2.

For convenience, Karmakar et al. (2000) provides a comprehensive table (Table 6.2) consisting of the values of a and b. In this simplified model, the rain intensity in the rain medium is considered not to vary along the path, that is, the rain intensity is homogeneous along the vertical path up to a height H_R. This height is assumed to be a level from which raindrops with a diameter larger than 0.1 mm fall and may be described as the physical rain height. The rain attenuation in the zenith direction is given by (refer to Equation 6.2)

$$A(z) = H_R a R^b \tag{6.5}$$

But if we look at equation 6.2, we see that the values of $\lambda(s)$ is replaced by aR^b. So it appears that the simple multiplication of H with aR^b would provide $A(z)$ values. But for the sake of clarity, it is always suggested to find the cumulative distribution of the brightness temperature values at the chosen probability levels to convert the brightness temperature to attenuation values. These attenuation values are to be matched with the rain rate values given by the cumulative distribution of rain rate at the same probability levels. In this connection Mawira et al (1981) provides the measured values at 17.6 GHz as

$$A_{17.6} = A^0_{17.6} + H\gamma_{17.6}$$

where $A^0_{17.6}$ represents the contribution to the total attenuation from sources other than rain showers. The values obtained for $A^0_{17.6}$ are higher than those to be expected from contributions due to water vapor or gaseous absorption and cloud attenuation. It was found that the brightness temperature after the rain shower was about 45k higher than the nominal level before. This probably caused by the effects of water on the antenna-feed. So, proper care has to be taken before going for any radiometric measurement.

It is to be noted that the physical rain height is not easily measurable, the simplest approximation being identified with 0°C isotherm height, that is, the rain height during rainy conditions is given for latitudes less than 36° by the relation (ITU 2001)

$$H_R = 3.0 + 0.028\varphi \tag{6.6}$$

where φ is the latitude in degrees.

For tropical latitudes, $\varphi < 30°$, it is proposed by Ajayi and Barbaliscia (1990) that H_R be multiplied by a path reduction factor r, which is given by

$$r = \frac{1}{1 + \frac{L_G}{L_o}} \tag{6.7}$$

where $L_o = 35 \exp(-0.015r)$.

It is to be noted that the path reduction factor r has to be evaluated for 0.01% of time in a year, as far as the reliability factor is concerned.

A time series of zenith attenuation at 22.235 and 31.4 GHz and corresponding rain rates measured at Kolkata (22° N), India, on 27 July, 1991 is presented (Karmakar et al. 2000; Figure 6.2). A scatterplot of attenuation at 22.235 and

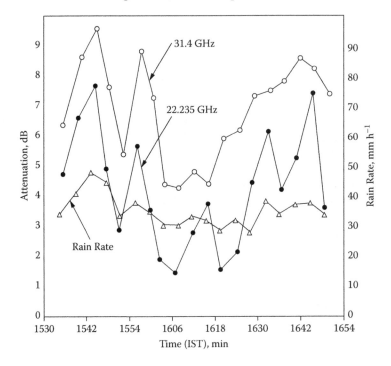

FIGURE 6.2
Time series of path attenuation at 22,235 and 31.4 GHz and corresponding rain rate over Kolkata (22° N), India 0n 27 July, 1991.

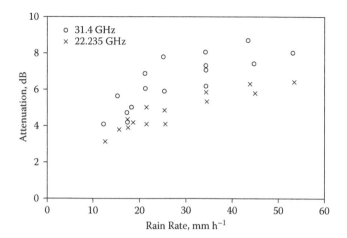

FIGURE 6.3

Scatter plot of rain attenuation at the frequencies as shown in Figure 6.2 and the corresponding rain rates for the same event.

31.4 GHz during the same year 1991 is presented in Figure 6.3. A regression analysis to a power law yielded the following best-fit equations:

$$A_{22}(dB) = 0.165R^{0.996} \quad (r^2 = 0.92) \tag{6.8}$$

$$A_{31}(dB) = 0.377R^{0.886} \quad (r^2 = 0.89) \tag{6.9}$$

These results were then compared with the specific attenuation as given by the following relationships (ITU-R 1994):

$$Y_{22}(dBkm^{-1}) = 0.087R^{1.05} \tag{6.10}$$

$$Y_{31}(dBkm^{-1}) = 0.184R^{0.99} \tag{6.11}$$

The rain attenuations at 31.4 are plotted against the rain attenuation at 22.235 GHz (Figure 6.4) together with the best-fit curve, given by the following equation:

$$A_{31} = 1.9A_{22}^{0.87} \tag{6.12}$$

with a correlation coefficient of 0.93. In comparison, the specific attenuation is related to each other by

$$Y_{31} = 1.8Y_{22}^{0.94} \tag{6.13}$$

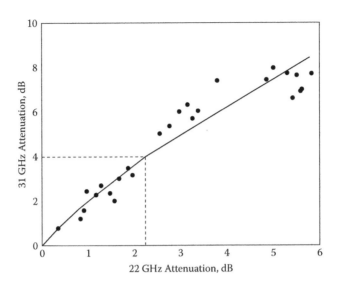

FIGURE 6.4
Scatter plot of attenuations at 22.235 GHz and those at 31.4 GHz during the year 1991.

The difference between these two expressions may originate from the fact that the volumes of the raining zone may not be identically illuminated. A comparison has also been made between the measured and calculated values of attenuations at 22.235 and 31.4 GHz taking into account all the drop size distributions cited in Table 6.1. It was found that the rain attenuations calculated with those drop size distributions differed significantly from the measured values, especially those at higher rain rates, except in case of the Joss et al.'s drizzle (JD) distribution, which fitted the measurements for rain rates up to 15 mm hr^{-1}. Table 6.3 presents the comparative values.

6.2.1 Rain Height

The nonuniform horizontal rain structure is accounted for by the use of a rain rate reduction factor to convert the physical path length to an effective path length. The simple vertical structure assumes that rainfall is uniform from the ground to rain height. The physical rain height is the level up to which the water drops with diameter larger than 0.1 mm present. However, the effective rain height may be obtained by analyzing the measured attenuation and point rainfall intensity data. Any nonuniformity of the vertical profile of rain is, in fact, integrated with time provided that all the water falling inside an ideal column ultimately reaches the ground. But it may so happen that a few raindrops remain aloft and then the radio wave propagation may not be affected. Moreover wind may also drive away the floating rain drops from the radio path.

TABLE 6.3

Comparisons of Attenuation at 22.235 and 31.4 GHz
(Joss et al. Distribution)

Rain Rate (mm/hr)	Attenuation (dB)	
	Calculated	Measured
Frequency = 22.235 GHz		
4	1.02	0.70
5	1.29	0.95
7	1.82	1.25
8	2.08	1.46
9	2.35	1.60
10	2.62	1.80
12	3.69	2.60
15	3.96	2.75
Frequency = 31.4 GHz		
4	2.20	1.25
5	2.79	1.65
7	3.95	2.40
8	4.55	2.70
9	5.14	3.50
10	5.74	3.90
12	6.95	3.90

But, in practice, the vertical nonuniformity is very unlikely to occur and it causes the effective rain height to seem higher than the physical rain height. The situation becomes more complex when horizontal nonuniformity occurs and is relevant in considering the global rain attenuation effect. At first approximation, the vertical nonuniformity may be disregarded in comparison to horizontal nonuniformity except for high elevation angle ($\varphi > 60°$). This is the reason for which the rain height may be used instead of rain thickness for both physical and effective measurements of rain attenuation.

The physical rain height is not easily measurable and the closest approximation for rain height is to consider the 0° isotherm, which is readily available from radiosonde data. But during rain, there lies a big difference between the two, which, in turn, depends on the types of rain. In warm rain, the physical rain height is lower than the 0° isotherm. In thunderstorm rain it is normally present well above the 0° isotherm and in stratiform rain, the 0° isotherm and the physical rain height become coincident. This happens especially in the cold season when the falling ice crystals melt below the 0° isotherm height (Ajayi and Barbaliscia 1990). The types of rain are detailed next for clarity.

6.2.2 Structure of Rain

Basically, there are three types of rains:

1. Stratiform rain—It is a type of rain that is widespread, and long lasting with low-medium intensity. These rains are generated by the sublimation–coalescence process and it involves the ice phase.
2. Warm rain—This type of rain generates by the condensation–coalescence process. It is typical to warmer seasons and starts well below the 0° isotherm. No ice phase is involved.
3. Thunderstorm rain—It is a convective rain and falls with high rain intensity along with thunder and lightning. These are generated with cumulous–nimbus cloud systems, which may extend up to 10 km or even higher due to strong updrafts.

6.2.3 Estimation of Rain Height

Referring to Figure 6.1 and Equation 6.5, we see that H_R is the rain height and the values of the constants a and b are obtainable at the desired frequencies. It is also noted that the JD distribution is well suited at Kolkata (22° N) at low rates, up to 15 mm hr^{-1}. So, keeping these in mind one can find out the measured rain height over the particular place of choice. Table 6.4 gives the value of rain height over Kolkata (Karmakar et al. 2000). It appears from Table 6.4 that the value is less than the 0° isotherm. Hence it may be concluded the type of rain present during the study period is warm rain.

TABLE 6.4

Values of Attenuation and Rain Height over Kolkata, India, for the Year 1991

Rain Intensity (mm/hr)	Rain Attenuation (dB)	Rain Height (km)
Frequency = 22.235 GHz		
0.5	0.85	2.73
0.8	1.36	2.69
10.8	1.84	2.67
13.5	2.30	2.65
Average	—	2.69
Frequency = 31.4 GHz		
0.5	1.55	2.39
0.8	2.48	2.34
10.8	3.35	2.31
13.5	4.20	2.29
Average	—	2.33

6.2.4 Variability of Rain Height

The several experimental results over the different parts of the world including temperate and tropical regions, reveal that the 0°C isotherm varies with several factors. In this context, Ajayi and Barbaliscia (1990) made a comprehensive study. The quantities h_{FM}, h_{FY}, h_{FS}, the mean values of the 0°C isotherm height in an average month, year, and summer half-year, respectively, and h_{FR}, the mean values for rainy conditions for various rain thresholds were taken into consideration. In the northern hemisphere the summer half-year includes the months from May to October but in the southern hemisphere it is from November to April. In the northern hemisphere the results were obtained from 3.4° to 46° N, and those in the southern hemisphere the latitude varied from 6.88° to 45.47° S. Tables 6.4 and 6.5 show the variation of the 0°C isotherm height in the northern hemisphere and in the southern hemisphere, respectively. It is observed from the tables that there lies a negligible difference between noon and midnight values of h_{FY} and h_{FS}. The diurnal variations over the temperate and tropical locations were found to be insignificant in comparison to monthly or seasonal variation. This suggests that one year data is adequate for studying the year-to-year variability of 0°C isotherm height over a particular place of choice. A high monthly variability is observed over Trapani, a temperate location (Figure 6.5). Similarly, in the tropical location like Minna, the monthly variation over a year is less than 5% (Figure 6.6).

The annual variation of h_{FR} during rainy conditions over tropical and temperate locations is presented in Figure 6.7 and Figure 6.8, respectively. The regression line for the temperate location is

$$h_{FR} = 2.125 + 0.009\,R$$

For the tropical location it is

$$h_{FR} = 4.67 + 0.006\,R$$

where R is rain rate in mm hr^{-1}.

These confirm the negligible dependence of rain rate on the 0°C isotherm height. But the determination of h_{FR} during rainy conditions is difficult. Here, it has been assumed that the significant rain occurs in the summer half-year. It has been observed that for the temperate location when summer rains are considered alone, the 0°C isotherm height appears to be almost independent of rain intensity at least up to 15 mm hr^{-1}. But beyond this there might be little dependence on rain intensity. So it is suggested to perform the rain height determination experiment within the limit of 10–15 mm hr^{-1}.

Now let us look back to Figure 6.5 and Figure 6.6. They show that quantity h_{FM} has an inverse variation with ground relative humidity. In temperate regions during the summer, there is an increase in the value of the 0°C

TABLE 6.5A

Variation of 0°C Isotherm Height in Northern Hemisphere (Location: Italy)

Station	Longitude (°E)	Latitude (°N)	h_{FY}(km)			h_{FS}(km)			h_{FR}(km) (10 mm/2h)		
			Noon	Midnight	Average	Noon	Midnight	Average	Noon	Midnight	Average
Roma	12.2	41.8	2.73	2.68	2.71	3.45	3.41	3.43	1.91	2.71	2.31
Brindisi	17.5	40.7	2.78	2.69	2.74	3.57	3.47	3.52	2.23	2.22	2.23
Cagliari	9.1	39.3	2.88	2.88	2.87	3.67	3.65	3.66	2.22	2.04	2.13
Trapani	12.5	37.9	3.23	3.23	3.16	3.97	3.91	3.94	2.21	2.28	2.25
Udine	13.2	46.0	2.37	2.37	2.34	3.09	3.02	3.06	2.10	2.51	2.31

TABLE 6.5B

Variation of 0°C Isotherm Height in Northern Hemisphere. The Parameter are showing the Average Values. (Location: Nigeria)

Station	Longitude (°E)	Latitude (°N)	h_{FY} (km)	h_{FS} (km)	h_{FR} (km)
Oshodi	3.4	6.5	4.74	4.70	4.54
Minna	6.5	9.6	4.81	4.75	4.68
Kano	8.5	12.1	4.79	4.74	4.79

isotherm height with a corresponding decrease in ground relative humidity. But in the tropics the case is reversed.

The variation of h_{FM} with ground temperature is also remarkable. The empirical relationship between 0°C isotherm height with ground temperature over a temperate location is found to be

$$h_{FM} = 0.186 + 0.142t \quad \text{(at noon)} \tag{6.14}$$

$$h_{FM} = 0.530 + 0.157t \quad \text{(at midnight)} \tag{6.15}$$

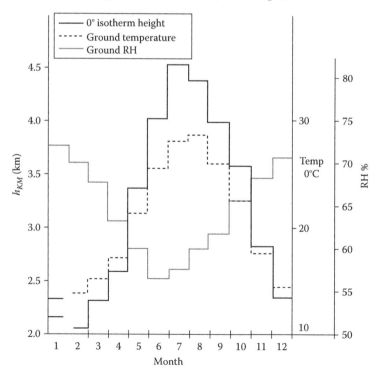

FIGURE 6.5

Monthly variation of 0° isotherm height, ground temperature, and relative humidity for Trapani, Italy.

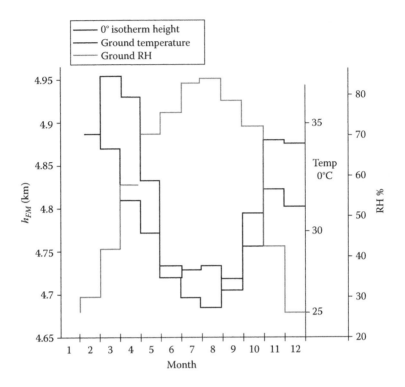

FIGURE 6.6
Monthly variation of 0° isotherm height, ground temperature, and relative humidity for Minna, Nigeria.

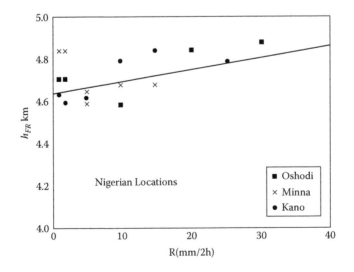

FIGURE 6.7
Variation of 0° isotherm height (km) with rain threshold for Italy.

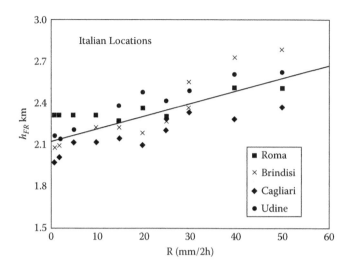

FIGURE 6.8
Variation of 0° isotherm height (km) with rain threshold for Nigeria.

Here h is the 0°C isotherm height in km and t is the ground temperature in degree Celsius (Figure 6.9). On the other hand, over a tropical location the relationship was found to be

$$h_{FM} = 4.0 + 0.026t \qquad (6.16)$$

This is presented in Figure 6.10.

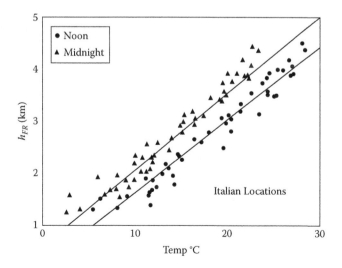

FIGURE 6.9
Variation of 0° isotherm height with ground temperature for Italy.

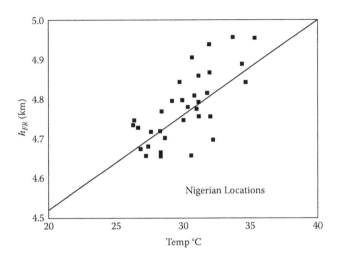

FIGURE 6.10
Variation of 0° isotherm height with ground temperature for Nigeria.

It is observed from Figures 6.9 and 6.10 that the 0°C isotherm height is relatively more sensitive with ground temperature in temperate locations than that of tropical locations. For the same ground temperature change, h_{FM} at the temperate location changes more than five times the amount that h_{FM} changes at the tropical locations. Figure 6.11 shows the variation of h_{FM} with ground temperature for summer half-year rainy events for the northern hemisphere and is given by

$$h_{FR} = -0.171 + 0.174t \qquad (6.17)$$

On the basis of the global scale, the variation of ground temperature is found to be the following (Ito 1989).

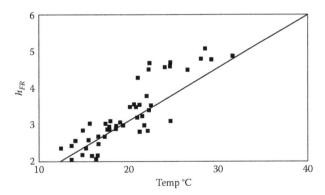

FIGURE 6.11
Variation of 0° isotherm height with ground temperature for summer rain in Northern hemisphere.

For the northern hemisphere,

$$t = 27.3°C; \quad \text{latitude } \varphi \leq 26° \tag{6.18a}$$

$$t = 27.3 - 0.50 \, (\varphi - 26); \quad \text{latitude } \varphi > 26° \tag{6.18b}$$

For the southern hemisphere,

$$t = 25.5; \quad \text{latitude } \varphi \leq 26° \tag{6.19a}$$

$$t = 25.5 - 0.642(\varphi - 26); \quad \text{latitude } \varphi > 26° \tag{6.19b}$$

Using Equations (6.17) and (6.18), the variation of 0°C rain height in the northern hemisphere is

$$h_{FR} = 4.6 - 0.084(\varphi - 26) \tag{6.20}$$

Figure 6.12 shows the variation of 0° isotherm with ground temperature for the southern hemisphere and the corresponding equation is

$$h_{FR} = 0.68 - 0.155t \tag{6.21}$$

Using Equations (6.19) and (6.21) we have

$$h_{FR} = 4.6 - 0.1(\varphi - 26) \tag{6.22}$$

Hence we have the 0°C isotherm as a function of latitude for the northern and southern hemispheres (refer to Equations 6.20 and 6.22). Both the representations look to be similar except for the difference in their slope, which might be due to the geographical origin.

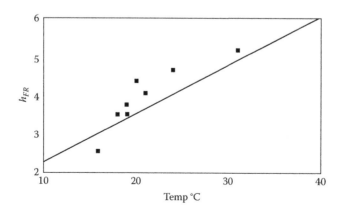

FIGURE 6.12
Variation of 0° isotherm height in rainy condition with temperature in January in the southern hemisphere.

6.3 ITU-R Rain Attenuation Model and Its Applicability

Several researchers were largely engaged in developing reliable techniques for the prediction of path rain attenuation since last decade. Several models have been proposed in this regard. These models are largely frequency and location dependent and were developed on the basis of satellite beacon measurements (Li and Yang 2006). However, the mostly accepted prediction procedure of calculating rain attenuation is based on the ITU-R (2003, 2005) model and Crane (1996) model. Incidentally, most of the rain attenuation models were based on the data collected from temperate regions (Singh et al. 2007). Rain in these regions is mostly of stratiform structure, which is generally light with relatively large rain cell diameters. But in the tropics, rain at times is from convective rain cells, with relatively small diameters often resulting in heavy downpours for short periods (Ramachandran and Kumar 2005; Mandeep and Allnutt 2007). In temperate regions, due to large rain cell diameters, the attenuation increases with a decrease in elevation angle. But in the tropical climate, for the same rain rate, the converse is found to be true (Pan et al. 2001; Pan and Allnutt 2004). In this context, Mandeep and Allnutt (2007) reported the results obtained from experimental data over the University Sains Malaysia (USM), Institute Technology Bandung (ITB), Ateneo de Manila University (AdMU), and University of South Pacific (USP). The exceedence curves in Figures 6.13 and 6.14 show that as the rain rate increases, the trend of the slope of the curve decreases gradually from a large negative value and then trend is reversed. The point at which the change occurs is referred to as the break point in the exceedence curve. This

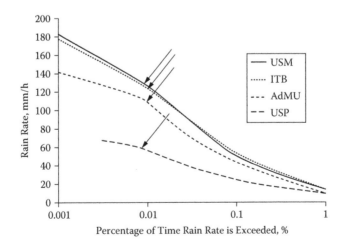

FIGURE 6.13
The exceedence curves show that as the rain rate increases, the trend of the slope of the curve decreases gradually from a large negative value, and then the trend is reversed.

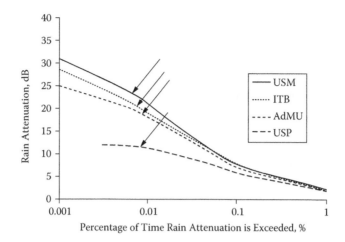

FIGURE 6.14
The exceedence curves show that as the rain rate increases, the trend of the slope of the curve decreases gradually from a large negative value, and then the trend is reversed.

occurs usually at high rain rates. In the tropics, when the cloud builds up, the water droplets are trapped in updrafts inside the cloud and are vertically transported. This enhances the coalescence of water particles resulting in convective heavy rain (Schumacher and Houze 2003). Thus any change in the trend of the exceedence curve signifies that the rain structure changes from stratiform rain to convective rain. To validate this idea, based on which the aforementioned discussions were made, the modified ITU-R model is proposed by Ramachandran and Kumar (2007).

This model closely follows the measured rain attenuation as reported by Mandeep and Allnutt (2007) for all the measurement sites throughout the entire percentage of time where rain attenuation is exceeded. The Flavin model (Flavin 1982) underestimates the measured rain attenuation values from 1% to 0.01% before overestimating the measured values from 0.01% to 0.001% of time rain attenuation is exceeded. Basically, this model is based on the ITU-R model pertaining to the information of rain cell size and 0° isotherm height. The Yamada (1987) model follows closely from 1% to 0.01% of time rain attenuation is exceeded but it overestimates from 0.01% to 0.001% of time rain attenuation is exceeded in all three sites except at ITB. The Yamada model underestimates the measured values from 1% to 0.003% of time rain attenuation is exceeded. This is due to the fact that the use of vertical path reduction factor for the models that provide a direct calculation of the attenuation values that calculate the probability of exceeding the specified attenuation was not considered. Hence we need to have a correct estimation of rain rate, path length, and the path reduction factor at a single probability level. In this context, at low rain rates the ITU-R model behaves well with the measured values but deviates considerably at higher rain rates from

TABLE 6.6A

Variation of 0°C Isotherm Height in the Southern Hemisphere (Location: Argentina)

Location	Longitude (°W)	Latitude (°S)	h_{FY} (km) 000 GMT	1200 GMT	Average	h_{FS} (km) 000 GMT	1200 GMT	Average
Salta	65.29	24.51	4.67	4.37	4.52	4.99	4.78	4.89
Resistencia	59.03	27.27	4.42	4.42	4.42	4.63	4.63	4.63
Cordoba	64.13	31.19	3.79	3.60	3.70	4.20	3.97	4.01
Ezeiza	58.32	34.49	3.38	3.32	3.35	3.83	3.81	3.82
SantaRosa	64.16	36.34	3.09	2.76	2.93	3.57	3.33	3.45
Cdte.Espora	62.10	38.40	2.89	2.68	2.79	3.38	3.24	3.31
Neuquen	67.59	38.57	2.89	2.58	2.74	3.46	3.13	3.30

0.01% to 0.001% of time rain attenuation is exceeded. So the ITU-R model needs proper modification, especially in the tropical zones where rain rates often reach very high values. However, in the next section we will discuss the procedure for calculating earth–space path attenuation based on the ITU-R model, which will be helpful in modifying this model as proposed by Ramachandran and Kumar (2007).

6.3.1 Procedure for Calculating Attenuation Based on the ITU-R Model

The prediction procedure provided here is considered to be valid up to 40 GHz and path length up to 60 km. The specific attenuation (dB/km) at $R_{0.01}$ is calculated by using the simple power law relationship as

$$\gamma_R = KR_{0.01}^\alpha \tag{6.23}$$

1. One has to obtain rain rate exceeded for 0.01% of time, with an integration time of 1 minute, from ITU-R Recommendation P.837. Values for coefficients k and α are determined as function of frequencies ranging from 1 to 1000 GHz and are presented graphically and in tabular form in ITU-R Recommendation (p. 838–843).

2. Compute the effective path length d_{eff}, by multiplying the actual path length d in kilometers by a distance factor r as

$$d_{eff} = dr$$

where $r = \frac{d_0}{(1+d)}, [d_0 = 35e^{-0.015R_{0.01}}]$

3. Thus an estimate of path attenuation exceeded for 0.01% of time is given by

$$A_{0.01} = \gamma_R d_{eff} = \gamma_R \, dr \, (dB)$$

TABLE 6.6B

Variation of 0°C Isotherm Height in Southern Hemisphere (Location: Tanzania)

Station	Longitude (°E)	Latitude (°S)	h_{FY}	h_{FS}
Dar es Salaam	39.7	6.88	4.65	4.76

4. For latitudes $\geq 30°$ north or south, the attenuation exceeded for other percentage of time p in the range 0.001% to 1% may be deduced from the power law as

$$\frac{A_p}{A_{0.01}} = 0.12 p^{-(0.546+0.043 \log_{10} p)}$$

where p is the probability of exceedence in percent. It comes out as 0.12 for $p = 1.00\%$; 0.39 for $p = 0.1\%$; 1.00 for $p = 0.01\%$, and 2.14 for $p = 0.001\%$.

5. For latitudes $\leq 30°$ north or south, the attenuation exceeded for other percentage of time p in the range 0.001% to 1% may be deduced from the power law as

$$\frac{A_p}{A_{0.01}} = 0.07 p^{-(0.855+0.139 \log_{10} p)}$$

where p is the probability of exceedence in percent. Here also the value of $\frac{A_p}{A_{0.01}}$ can be determined for several values of p.

6. If worst month statistics are required then we are to calculate the annual time percentages p corresponding to worst month time percentages p_w using climatic information given in recommendation ITU-R Recommendation (p. 841).

The annual rain outage probability is translated to worst month rain outage probability as (Lehpamer 2005)

$$p_w = 2.85 \, p_\alpha^{0.87}$$

6.3.2 Procedure for Modifying the ITU-R Model Applicable for Tropics

It has already been discussed that in the tropics the rain structure frequently changes from stratiform to convective. Hence there lies a chance to intercept more rain cells in the slant path. This needs to introduce a correction factor C_f to calculate the attenuation for paths with elevation angles less than 60° (Ramachandran and Kumar 2007). The attenuation can be calculated as

$$A_B = kR_B^\alpha L_E C_f$$

Here A_B is the break point attenuation and the correction factor C_f is given by

$$C_f = -0.002\theta^2 + 0.175\,\theta - 2.3$$

The percent exceedence at the break point attenuation has to be calculated. Ramachandran and Kumar proposed that for tropical locations the value of $p_{0.01}$ is 0.021%. According to ITU-R Recommendation P.618-8 (2003), the attenuation $A_{0.01}$ is calculated from $R_{0.01}$. But in this modified model $R_{0.01}$ is used to calculate $A_{0.021}$. In the modification proposed, p in the ITU-R model replaced by $p - 0.011$ so that when $p = 0.021\%$, $A_{0.021} = A_B$, the break point attenuation. The modified expression is

$$A_p = A_B \left[\frac{p - 0.011}{0.01} \right]^{-[0.655 + \ln(p - 0.011) - 0.045\ln(A_B) - \beta(0.989 - p)sin\theta]} \qquad \text{dB, for } 0.021 \le p < 1$$

$$(6.24)$$

As discussed earlier, in the tropics when the rain rate increases and approaches the break point the corresponding rain structure changes. It has also been indicated by radar observation that the convective rain cells are surrounded by stratiform rain (Schumacher and Houze 2003). So, in attenuation estimation the effective path length in the different regions should be considered. Ramachandran and Kumar (2007) show that the prediction model for attenuation should be related to the magnitude of attenuation at the break point, and beyond this the rain is assumed to be convective and droplets are spherical. However the expression for attenuation beyond this break point can be viewed as modification to the ITU-R model as

$$A_p = A_B \left(\frac{p}{0.021} \right)^{-0.5\{0.655 + 0.033\ \ln p^2 - 0.03\ln p - 0.045\ln A_B - \beta(0.989 - p)\sin\theta\}} \qquad \text{dB, for } p \le 0.021$$

$$(6.25)$$

Equation 6.24 gives a gradual increase in attenuation with increasing rain rate and Equation 6.25 shows the attenuation tending to saturation.

6.4 Raindrop Size Distribution in the Tropics

The process of raindrop formation, growth, transformation, and decay occur on a microphysical scale within a large cloud scale environment. Each process such as condensation growth/evaporation or collision/coalescence leaves a signature on the drop size distribution (hereafter DSD) of the rain event.

Hence analysis of the form of the DSD, its temporal and rain-rate-dependent evolution at the surface and also aloft is essential in understanding the process of rainfall formation. Recent DSD studies have focused on the differences between convective and stratiform rain, their differing characteristics and the physics of their formation. Houghton (1968) pointed out that the primary precipitation growth process in convective precipitation is a collection of cloud water by precipitation particles in strong local updraft cores. As the parcels of air in the convective updrafts rise out of the boundary layer and reach the upper troposphere, they broaden and flatten as a result of decreasing vertical velocity of the parcel and on reaching their level of neutral buoyancy, spread out and amalgamate to form a large horizontal area, which we identify as a stratiform region on the radar (Yuter and Houze 1995). Under these conditions, the primary precipitation growth process is vapor deposition on ice particles formed earlier in the convective updrafts, but left aloft as the convective drafts weaken. It is a slow process, with particles always settling downward.

It was observed by Roy et al. (2005) that during the rain event, at low rain rates, the convective phase of the rainfall event was marked by DSD spectra that have greater population of small droplets as compared to stratiform DSDs at the same rain rates. At higher rain rates, the convective regime is characterized by narrow spectra centered at higher diameters. At the transition region between convective and stratiform spectra, mixed large and small drop spectra are observed.

Similar variation was also observed in the averaged drop spectra. In addition, the averaged spectra also reveal an equilibrium distribution of the drop population in DSDs at higher rain rates (>39/hr) for diameter range (>1.91 mm) corresponding to nearly constant values of the slope of the distribution, the intercept, and the mean mass diameter. The value of the shape parameter, which for small rain rates varies the same as the slope parameter, starts to increase with increasing rain rate as the other two parameters of the gamma distribution approach a constant value corresponding to equilibrium shape. The value of the intercept parameter is highest for low to moderate convective rainfall and decreases as the rain rate increases. With the availability of reliable disdrometer observations of rainfall DSD, this became a major tool for identifying cloud types from rainfall received on the ground. Most studies of rainfall DSDs have observed a sudden decrease in the value of the intercept parameter N_0, for exponential and gamma DSD in association with transition of rainfall type from convective to stratiform. This result was first reported by Waldvogel (1974) and has since been consistently observed for tropical rainfall by various authors (Heinrich et al. 1996; Maki et al. 2001). They demonstrate a clear relationship between the riming process in clouds and N_0 of the raindrop spectra, all of which change dramatically as riming increases, at times without a corresponding change in the rain rate. On the other hand, Stewart et al. (1984) have demonstrated the presence of large raindrops associated with the aggregation mechanism above the bright band in stratiform rainfall regions. Since riming (an indication of updrafts and convection) is the main process determining the form of the DSD in convective clouds, and aggregation is the most important process in stratiform DSD formation (Atlas and Ulbrich 2000), one may logically associate small drop DSD spectra (large N_0 values) with convective clouds and large drop spectra (small N_0) with stratiform mode of DSD formation at the same rain rate. Following the physical arguments of Waldvogel (1974), Tokay and Short (1996) found that the relation $N_0 = 4 \times 109 \times R^{-4.3}$ was a good threshold to distinguish between convective and stratiform precipitation in tropical rainfall. Tokay et al. (1999) compared the disdrometer based algorithm with the 915 MHz wind-profiler-based rainfall separation method developed by Williams et al. (1995) and found reasonable agreement between the two methods for tropical rainfall.

6.4.1 Case Study over Southern Part of India

Roy et al. (2005) reported that the raindrop DSD was measured with a RD-80 model disdrometer, which is an advanced version of the RD-69 model developed by Joss and Waldvogel (1967). A RD-80 model disdrometer (Distromet Ltd., Switzerland) was installed at Cuddalore (11.43° N, 79.49° E) under the Indo–US project in April 2002 to measure the rain rate along with the DSD of tropical continental rainfall.

6.4.1.1 Case 1

A squall line rain event of September 15, 2002, comprising both heavy and light showers over the station 18:31 IST to 21:14 IST were analyzed (Roy et al. 2005). (IST signifies the Indian Standard Time, which is five and a half hours ahead of UTC.) The accumulation during the event was 43.6 mm and the peak rain rate associated with the heavy rainfall regime, measured by the disdrometer is 101.8 mm/hr. The squall line was observed in the radar pictures as a northeast–southwest oriented band of high reflectivity that moved in a southeasterly direction over the observation site, before turning and moving south-westward and then finally losing its distinctive form and ending its life as a cloud cluster.

Figure 6.15 displays a line diagram of R versus N_0. It may be observed from the same figure that there is a clear separation of the N_0 value into two groups, one corresponding to continuous, low intensity long period rainfall and the other, the high intensity, short period rainfall. Hence the Tokay and Short (1996) classification may be considered applicable to this particular rain event. However it may be noted, that the value of the $N_0 - R$ relationship (as indicated in Figure 6.15 by the heavy gray line) that separates the entire event into two rainfall types is given by $5.8 \times 10^7 R^{-4.45}$, which is about hundred times lower than the value obtained by Tokay and Short. This may be because of

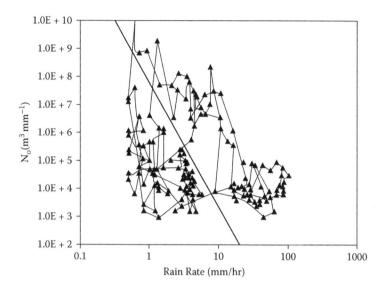

FIGURE 6.15
The value of N_0–R relationship, as indicated by the heavy grey line, that separates the entire event into two rainfall types.

the fact that they employed the third, fourth, and sixth moments of the DSD in determining the parameters of the gamma distribution; whereas in the present study, the second, fourth, and sixth moments of the DSD have been employed. All three parameters of the gamma distribution are very sensitive to the moments of the raindrop spectra. The relation between the intercept parameter and rain rate is therefore expected to be different than the study made by Tokay and Short.

The corresponding temporal variation of the classification scheme may be observed in Figure 6.16a. The columns indicate convective rainfall. Since stratiform rain rates are very low compared to convective rain rates, a scale change has been effected for the stratiform rain rates as indicated by the line with solid circles. The convective classification was predominant in the initial light rain, as well as the heavier showers that followed until 20:09 IST, when the leading convective line and the transition region was directly over the station. The rainfall became predominantly stratiform thereafter, up to 21:14 IST at the end of the rainfall event corresponding to the presence of the secondary maximum in reflectivity over the station.

6.4.1.2 Case 2

A widespread cloud patch was observed in the neighborhood of the station on September 18, 2002. The entire low reflectivity cloud mass was moving in a south-eastward direction, while short lived very intense

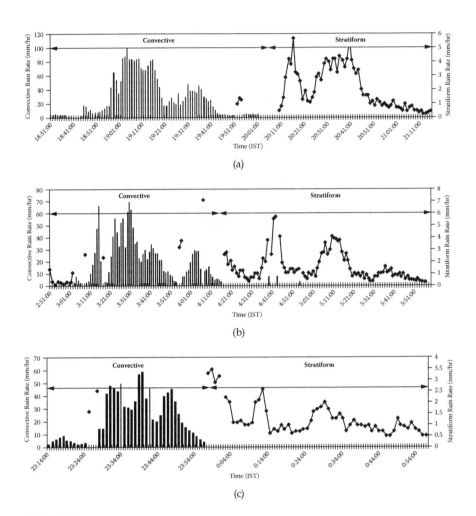

FIGURE 6.16

(a) Temporal variation of the classification scheme may be observed. The columns indicate convective rainfall. (b) A part of the total time series of convective cell, when it was directly over the station at 03:30 IST. There was very little rainfall (<0.5 mm/hr) in the preceding two hours and this rainfall spell started at 02:51 IST. At 04:15 IST, the rainfall changed to stratiform and continued to be so until the end of the rainfall period. (c) The convective classification was predominant in the initial light rain until 23:27 IST followed by heavier showers until 23:57 IST, when the leading convective line was directly over the station.

convective cells formed and decayed in this cloud cluster. The actual rainfall extended intermittently from 20:30 IST of September 17 to 05:37 IST of September 18.

However, in Figure 6.16b, a part of the total time series starting when a convective cell was directly over the station has been displayed at 03:30 IST. There was very little rainfall (<0.5 mm/hr) in the preceding two hours and this rainfall spell started at 02:51 IST with predominantly convective rain on the ground corresponding to the high reflectivity convective cell overhead. At 04:15 IST the rainfall changed to stratiform and continued to be so till the end of the rainfall period.

6.4.1.3 Case 3

A northwest southeast oriented squall line on the night of October 10, 2002, was observed in the neighborhood of the station moving in a north-easterly direction. Rainfall occurred for a total of 1 hour 44 minutes from 23:14 IST of October 10 to 00:58 IST of October 11. Peak rainfall was 62.5 mm/hr while the total accumulated rainfall was 18.23 mm. As displayed in Figure 6.16c, the convective classification was predominant in the initial light rain until 23:27 IST followed by heavier showers until 23:57 IST when the leading convective line was directly over the station. Following the convective line, the separation of the transition and stratiform regions are not evident unlike the first case, and the trailing low reflectivity region was observed to give predominantly stratiform rainfall on the ground up to the end of the rainfall period.

The aforementioned classification scheme has been applied to all kinds of rainfall events during the season, both from single cumulus cells as well as multicellular complex cloud structures. Although, the jump in the value of N_0 is not so unambiguous in all rainfall events of this study, it is observed that this N_0–R relation best separates the rainfall DSD of most rainfall events into contiguous groups, identified here as convective and stratiform (below the bold line in Figure 6.15). Minute-to-minute variations in classification can be occasionally seen in the time series revealing the empirical nature of the method. However the classification is quite stable. A 10% change in the slope or intercept of the N_0–R relation brings about only a 2%–3% change in the classification scheme. The different phases of the rainfall in the studied events have been broadly categorized as convective and stratiform type using the modified Tokay and Short algorithm in the previous section, given the complexity of structures observed in tropical precipitation. This also gives substantial systematic temporal variation of the DSD parameters often observed within a rainfall event, even for the same rain type and rain rate, depending upon the DSD of the particular location in the overall rain event.

6.4.2 Case Study over Northern Part of India

Raindrop size distribution measurements have been carried out at different locations of India as reported by several authors. Incidentally all of them prescribed three models in terms of lognormal function pertaining to three locations. Generally the lognormal N(D) was described in Chapter 1, Section 1.8.1.4. In this context, Maitra (2000) reported the path averaged DSD values from rain attenuation measurements at 94 GHz and 0.63 μm at Kolkata

(22° N, 88° E) for a year, which yielded the following models for the distribution parameters:

$$N_T = 546R^{0.469}$$

$$\mu = -0.538 + 0.017 \ln R$$

$$\sigma^2 = 0.0689 + 0.076 \ln R$$

where R is the rain rate in mm/hr.

Timothi et al. (1995) reported DSD measurements over a 2-year period with an optical disdrometer at Guwahati (26° N, 91° E), which yielded the distribution parameters as

$$N_T = 180\ R^{0.39}$$

$$\log \mu = 1.12 + 0.18 \ln R$$

$$\sigma^2 = 0.48 - 0.03R$$

The third model considered is for the location of Dehradun (30° N, 78° E) from which the distribution parameters obtained from observations over a year with an impact type disdrometer are expressed as (Verma and Jha 1996)

$$NT = 169\ R^{0.294}$$

$$\mu = -0.056 + 0.131 \ln R$$

$$\sigma^2 = 0.3 - 0.024 \ln R$$

According to Maitra (2004), for Kolkata and Dehradun, no classification of rain types was made and the DSD models are supposed to be average models for shower to thunderstorm rains, whereas for Guwahati the DSD model is mentioned to be appropriate for thunderstorms.

References

Ajayi, G. O., Barbaliscia, F., 1990, Prediction of attenuation due to rain: Characteristics of the 0° isotherm in temperate and tropical climates, *International Journal of Satellite Communications*, 8, 187–196.

Atlas, D., Ulbrich, C. W., 2000, An observationally based conceptual model of warm oceanic convective rain in the tropics, *Journal of Applied Meteorology*, 39, 2165–2181.

Bodtman, W. E., Ruthroff, C. L., 1974 Rain attenuation on short radio paths: Theory, experiment and design, *Bell System Tech. Journal*, 53, 1329–1349.

Crane, R. K., 1996, Electromagnetic wave propagation through rain, New York, John-Wiley and Sons.

Donnadieu, G., 1982, Observation de deux changements des spectres des gouttes de pluie dansune averse de nuages stratiformes. *J. Rech. Atmos.*, 16, 35–45.

Flavin, R. K., 1982, Rain attenuation considerations for satellite paths in Australia. *Australian Telecom Research*, 23, 2, 47–55.

Gunn, K. L., Kiner, G. D., 1949, The terminal velocity of fall for water droplets in stagnant air, *Journal of Meteorology*, 6, 243–248.

Harden, B. N., Norbury, J. R., White, J. K., 1977, Measurement of rain fall for studies of millimetric radio attenuation, *IEE. J. Microwaves, Opt and Acoustics*, 1, 6, 197–202.

Henrich, W., Joss, J., Waldvogel, A., 1996, Rain drop size distributions and the radar bright band, *Journal of Applied Meteorology*, 35, 1688–1701.

Houghton, H. G., 1968, On precipitation mechanism and their artificial modification, *Journal of Applied Meteorology*, 7, 851–859.

Ito, S., 1989, Dependence of the 0° isotherm height on temperature at ground level in rain, *Transactions of IEICE*, 72, 98–100.

ITU-R, 1974, Accuracy of frequency measurements at stations for international monitoring, Recomm. ITU-R,SM. 377–383, 1.

ITU-R, 2001, Recommendation, Propagation data and prediction methods required for design of terrestrial Line-of-sight systems, 623–633.

ITU-R, 2003, Recommendation, Propagation data and prediction methods required for design of erath space telecommunication systems, 618–628.

ITU-R, 2005, Specific attenuation model for rain for use in prediction methods, RECOMMENDATION ITU-R P, 838–843.

Joss, J. and Waldvogel, A., 1967, Ein spektrograph fur Niederschlag tropfen mit automatish auswertung. *Pure Appl. Geophys.*, 60, 240–246.

Karmakar, P. K., Chattopadhyay, S., Sen, A. K., Gibbins, C. J., 2000, Radiometric measurements of rain attenuation and estimation of rain height, *Indian Journal of Radio and Space Physics*, 29, 95–101.

Karmakar, P. K., Bera, R., Tarafdar, G., Mitra, A., Sen, A. K., 1991, Millimeter wave attenuation in relation to rain rate distribution over a tropical station, *International Journal of Infrared & Millimeter Waves (USA)*, 12, 11, 1333–1348.

Lehpamer, H., 2005, *Microwave transmission networks*, New Delhi: Tata McGraw- Hill.

Maitra, A., 2000, Three parameter raindrop size modeling at a tropical location, *Electronics Letter*, 36, 906–907.

Maitra, A., 2004, Rain attenuation modeling from measurements of rain drop size distribution in the Indian region, *IEEE Antennas and Wireless Propagation Letters*, 3, 180–181.

Maki, M., Keenan, T. D., Sasaki, Y., Nakamura, K., 2001, Characteristics of the rain drop size distribution in tropical continental squall lines observed in Darwin, Australia, *Journal of Applied Meteorology*, 40, 1393–1412.

Mandeep, J. S., Allnutt, J. E., 2007, Rain attenuation prediction at ku-band in South East Asia countries, *Progress in Electromagnetic Research*, 76, 65–74.

Mawira, A., Neessen, J., Zelders, F., 1981, Estimation of the effective spatial extent of rain showers from measurements by a radiometer and a rain gauge net work: Prediction of rain attenuation on slant path, *2nd ICAP, IEE Conf. Proc.* 133–136.

Pan, Q. W., Allnutt, J. E., 2004, 12 GHz fade durations and intervals in the tropics, *IEEE Transactions on Antenna and Propagation*, 52, 3, 693–701.

Pan, Q. W., Bryant, G. H., McMahon, J., Allnutt, J. E., Haidara, F., 2001, High elevation angle satellite to Earth 12 GHz propagation measurements in the tropics, *International Journal of Satellite Communications*, 19, 4, 363–384.

Ramachandran, V., Kumar, V., 2005, Invariance of accumulation time factor of ku-band signals in the tropics, *Journal of Electromagnetic Waves and Application*, 19, 11, 1501–1509.

Ramachandran, V., Kumar, V., 2007, Modified rain attenuation model for tropical regions for Ku band signals, *International Journal of Satellite Communication*, 25, 53–67.

Roy, S. S., Datta, R. K., Bhatia, R. C., Sharma, A. K., 2005, Drop size distributions of tropical rain over south India, *GEOFIZIKA*, 22, 105–131.

Schumacher, C., Houze, R. A., 2003, Stratiform rain in the tropics as seen by the TRMM precipitation radar, *American Meteorological Society*, 16, 1739–1756.

Singh, M. S. J., Hassan, S. I. S., Ain, M. F., Igarashi, K., Tanaka, K., Iida, M., 2007, Rain attenuation model for S.E. Asia countries, *IET Electronic Letters*, 43, 2, 75–77.

Stewart, R. E., Marwitz, J. D., Pace, J. C., Carbone, R. E., 1984, Characteristics through the melting layer of stratiform clouds, *Journal of Atmospheric Science*, 41, 3227–3237.

Timothi, K. I., Sharma, S., Devi, M., Barbara, A. K., 1995, Model for estimating rain attenuation at frequencies in range 6-30 GHz, *Electronics Letter*, 31, 1505–1506.

Tokay, A., Short, D. A., 1996, Evidence from tropical rain drop spectra of the origin of rain from stratiform versus convective clouds, *Journal of Applied Meteorology*, 35, 355–371.

Tokay, A., Short, D. A., Ecklund, W. L., Gage, K. S., 1999, Tropical rainfall associated with convective and stratiform clouds: Intercomparison of disdrometer and profiler measurements, *Journal of Applied Meteorology*, 38, 302–320.

Verma, A. K., Jha, K. K., 1996, Rain drop size distribution model for Indian climate, *Indian Journal of Radio and Space Physics*, 25, 15–21.

Waldvogel, A., 1974, The N_0 jump in rain drop spectra, *Journal of Atmospheric Science*, 31, 1067–1078.

Watson, P. A., Leitao, M. J., Turney, O., Sengupta, N., 1982, Development of a climatic map of rainfall attenuation for Europe, School of Electrical and Electronic Engineering, University of Bradford, *Reprt 327* (under ESA/ESTEC), contract No.4162/79/Nl/DG(SC).

Watson, P. A., Leitao, M. J., Sathiascelan,V., Gunes, M., Baptista, J. P. V. P., Turney, O., Brurraard, G., 1987, Prediction of attenuation on satellite—earth links in the European region, *IEE Proc.* 134, Pt. F, 6, 583–596.

Williams, C. R., Ecklund, W. L., Gage, K. S., 1995, Classification of precipitating clouds in the tropics using 915 MHz wind profiler, *Journal of Atmospheric and Oceanic Technology*, 12, 996–1012.

Yamada, M. Karasawa, Y., Yasunaga, M., 1987, An improved prediction method for rain attenuation in satellite communication operating at 10–20 GHz, Radio Sc., 22, 1053–1062.

Yang, P., 2006, The permittivity based on electromagnetic wave attenuation for rain medium and its applications, *Journal of Electromagnetic Waves and Applications*, 20, 15, 2231–2238.

Yuter, S. E., Houze, R. A. Jr., 1995, Three dimensional kinematic and microphysical evolution of Florida cumulonimbus, Part III: Vertical mass transport, mass divergence, and synthesis, *Monthly Weather Review*, 123, 1964–1983.

7

Attenuation by Hydrometeors Other than Rain

7.1 Snow

Snow is a mixture of air and ice crystals. The information about size distribution of snowflakes or liquid water content is very scanty. The international commission on snow and ice in the year 1949 classified the solid snow precipitation into ten categories. They are (1) plates, (2) stellar crystals, (3) columns, (4) needles, (5) spatial dendrites, (6) capped columns, (7) irregular particles, (8) graupel (soft hail), (9) ice pallets, and (10) hail. According to their sizes, they are categorized into five groups ranging from very small (0–0.49 mm) to very large (>4 mm). Among these the mean diameter of the largest crystal is presented in Table 7.1 for air temperature between –8°C and –15°C. It was also observed by Nakaya and Terada on Mt. Tokati that for plane dendrite particles, the thickness remains about 11 micrometers and independent of particle size. The mean density of the graupel was 0.125 g/cm³ with a maximum value 0.3 g/cm³. This information, however, was not sufficient for the wave propagation studies.

However, the snow mass concentration may be converted to the equivalent rainfall rate according to the relation given by

$$R_e = V \rho_{snow} \tag{7.1}$$

where ρ is the density of snow, V is the velocity of fall of snow, and R_e is the equivalent rain rate (mm/hr). Instantaneous rates of rainfall may vary considerably within one hour. This idea may be made clear by citing an example. The 0.06 inch precipitation during a clock hour may seem to be accumulated in 30 minutes at a rate of 0.12 inch/hr or in 20 minutes at the rate 0.18 inch/hr or linearly over the entire hour. However, the distribution curves (Figure 7.1) from instantaneous (1-minute period) and hourly observations for the same station show how the instantaneous rate distributes itself around the hourly rate. These instantaneous curves may be used to break down the long-term hourly rates into shorter instantaneous segments and the segments then compounded into a cumulative yearly distribution curve. Figure 7.2 shows

TABLE 7.1

Diameter, Mass, and Fall Velocity of Snow

Snow Crystal	Diameter (mm)	Mass (mg)	Velocity (cm/sec)
Needle	1.53	0.004	50
Plane dendrite	3.26	0.043	31
Spatial dendrite	4.15	0.146	57
Powder snow	2.15	0.046	50
Rimed crystal	2.45	0.176	100
Graupel	2.13	0.800	180

cumulative curves and similar curves for two hourly and half-hourly precip-
itations rates. A comparison of the frequency curves indicates that it makes
little difference in the annual frequency for rainfall rates between 0.06 to
0.18 mm hr^{-1}. The annual frequency of the instantaneous rate is approximately 90% of the mean hourly rate 0.06 inch h^{-1}; 94% for 0.12 in h^{-1} and 100%
for 0.18 in h^{-1}. Table 7.2 shows hourly frequencies over different stations all

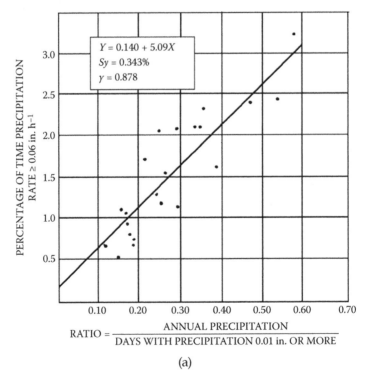

(a)

FIGURE 7.1
Correlation between annual probability of hourly precipitation equal to or exceeding (a) 0.06
in.h^{-1} (b) 0.12 in.h^{-1} (c) 0.18 in.h^{-1} and usually available precipitation data.

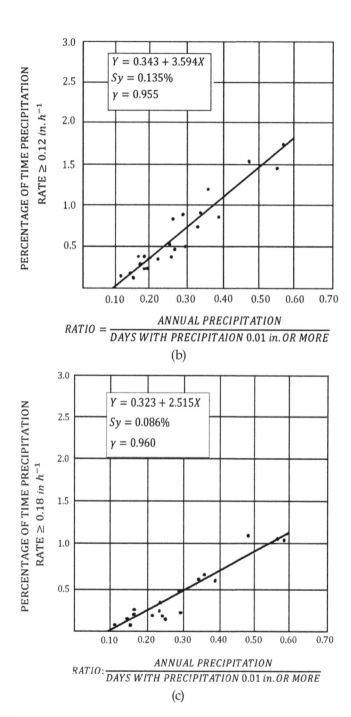

$$Y = 0.343 + 3.594X$$
$$Sy = 0.135\%$$
$$\gamma = 0.955$$

$$RATIO = \frac{ANNUAL\ PRECIPITATION}{DAYS\ WITH\ PRECIPITAION\ 0.01\ in.\ OR\ MORE}$$

(b)

$$Y = 0.323 + 2.515X$$
$$Sy = 0.086\%$$
$$\gamma = 0.960$$

$$RATIO: \frac{ANNUAL\ PRECIPITATION}{DAYS\ WITH\ PRECIPITATION\ 0.01\ in.\ OR\ MORE}$$

(c)

FIGURE 7.1
(Continued)

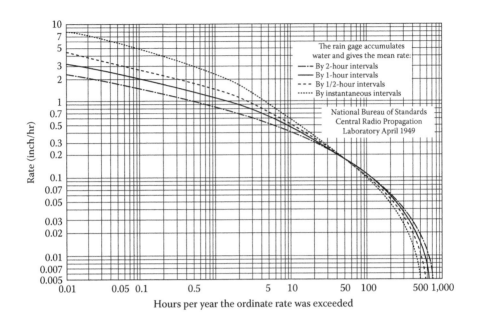

FIGURE 7.2
Cumulative distribution snow- and rainfall rate in Washington, DC. The 1-hour curve is based on longtime data. The other three curves are derived from the 1-hour curve.

over the world (Bussy 1950). Table 7.3 shows the probability of occurrence of given instantaneous rates of precipitation up to an exceeding 7.5 inch/hr for New Orleans. Incidentally New Orleans is a representative station in the area of heavy rainfall and one for which the frequencies of clock hourly rates based on 30 years of data are available. This clock hourly rates were converted to instantaneous frequencies by Bussey (1950).

Table 7.4 gives the approximate frequency with which given instantaneous rates of precipitation upto 1.5 inch/hr are equaled or exceeded in four climatic states of the United States of America. Data taken for Portland and Washington D.C are representative of conditions found on north-western and mid-eastern coasts respectively. New Orleans represents an area of heavy rainfall frequently under the influence of tropical maritime air and Oklahoma city represents the great plains. However, an empirical relationship for particle size distribution may be cited as

$$N_D = 0.025 R^{-0.94} \exp(-2.29 R^{-0.45} D) \tag{7.2}$$

where
N_D = number of drops per cc per centimeter size range
R = snow rate (mm/hr) of equivalent liquid water
D = diameter (cm) of a sphere of water formed

TABLE 7.2

Percentage of Time during an Average Year in Which Clock Hourly and Instantaneous Rates of Precipitation Equal or Exceed 0.12 and 0.18 in. hr^{-1} at Selected Stations

Station	Average Annual Precipitation (in.)	Number of Days with Measurable Precipitation	Clock Hourly Rate			Instantaneous Rates		
			0.06 (%)	0.12 (%)	0.18 (%)	0.06 (%)	0.12 (%)	0.18 (%)
Athens 37°30' N, 23°43' E	15.70	98	0.94	0.24	0.08	0.85	0.23	0.08
Berlin 52°30' N, 13°25' E	22.88	169	0.84	0.16	0.03	0.76	0.15	0.03
Dublin 53°20' N, 6°15' W	27.32	218	0.78	0.13	0.01	0.70	0.12	0.01
London 51°25' N, 0°20' E	24.47	167	0.89	0.20	0.06	0.80	0.19	0.06
Moscow 55°45' N, 37°37' E	24.13	132	1.05	0.30	0.13	0.95	0.29	0.13
Paris 48°52' N, 2°20' E	22.62	160	0.84	0.16	0.03	0.76	0.15	0.03
Rome 41°45' N, 12°15' E	26.70	105	1.45	0.59	0.33	1.31	0.55	0.03
Tokyo 35°41' N, 139°46' E	61.40	149	2.22	1.13	0.72	2.00	1.06	0.72
Warsaw 50°14' N, 21°00' E	22.21	164	0.84	0.16	0.03	0.76	0.15	0.03
Washington 38°55' N, 77°00' W	42.20	124	2.11	0.90	0.60	1.90	0.85	0.60

TABLE 7.3

Frequency with which instantaneous rates of precipitation
equal or exceed the indicated rates in New Orleans (Ref: R.D.
Fletcher, 1956 , Trans., Am. Geophys. Union, 31, P: 344)

Instantaneous Rate (inch /hr)	Frequency(%)	Frequency (h /yr)	Occurrence Probability
0.06	2.16	189	1 in 46
0.18	1.08	95	1 in 92
0.40	0.56	49	1 in 179
0.80	0.37	32	1 in 274
1.50	0.21	18	1 in 487
3.00	0.044	3.85	1 in 2275

The experimental results obtained by different workers in the frequency range 10–140 GHz may be summarized as

1. Attenuation coefficient increases approximately linearly with snow-fall rate.
2. Wet snow produces higher attenuation than dry snow.
3. Attenuation increases with frequency.

Moreover, the experimental results of snow attenuation by Nemarich in Vermont during 1981 at 96,140 and 225 GHz show that the attenuation at 225 GHz increases more rapidly with snow concentration and least with 96 GHz. It also shows that attenuation is more within the temperature range 0°C to 1.5°C than those within the range −0.5°C to 0°C. This predominantly shows that the higher is the temperature larger is the attenuation. This might happen due to larger amount of water content.

TABLE 7.4

Approximate percentage of time during an average tear when given instantaneous rates of precipitation are equaled or exceeded at four stations in the United States.

Station	percentage(%) of precipitation rates (inch/hr)								
	0.04	0.06	0.08	0.12	0.18	0.20	0.40	0.80	1.50
New Orleans	2.96	2.16	1.74	1.44	1.08	0.92	0.56	0.37	0.21
Oklahoma City	2.04	1.49	1.23	0.81	0.57	0.46	0.20	0.15	0.11
Washington D.C.	2.58	1.90	1.36	0.84	0.60	0.48	0.17	0.07	0.024
Portland, Oregon	1.84	1.34	1.090	0.57	0.35	0.28	0.11	0.05	0.016

7.2 Hail

The meteorological terminology of hail is mostly defined as ice particles with a diameter of 5 mm or more. Smaller particles are called ice pellets. Ice pellets are classified as sleet consisting of transparent and globular grains of ice. Hail is of the same type of sleet but, instead, it would be translucent. Sometimes larger hail has a diameter greater than 2 cm. Larger hail is formed only in well-developed cumulonimbus clouds when moisture is carried above about 30,000 ft. Smaller hail and soft hail are considered to be the essential features of all the thunderstorms. However, ice particles less than 1 cm diameter remain aloft in the core of the thunderstorm. These are likely to melt completely before reaching the ground. Therefore, although the occurrence of thunderstorm in tropics and subtropics are very frequent, hail is rarely found on the ground in tropics. Hailstorms are found to occur over mid latitude mountains and adjacent areas for 5 to 10 times in a year whereas the occurrence of thunderstorm is approximately 40 to 50 times in a year. Foster (1961) summarizes the physical properties of hail along with microphysics of hailstone growth in detail.

Hail may significantly affect the propagation of signals above few GHz. The empirical relationship in connection with number density of hailstones is given by

$$N_D = 31\exp(-3.09D) \qquad (7.3)$$

where D(cm) is the diameter of hailstone and N_D is the number of hailstones per c.c. with diameters varying between D and $D + \Delta D (\Delta D = 0.32$ cm). This distribution, however, is used to find out attenuation for a 15 GHz signal. It is found that the attenuation coefficient is approximately about 5 dB/km considering the maximum diameter of a hailstone of about 2.89 cm. However, the occurrence of hail has a large impact on aircrafts. Analyses during 1951 to 1959 over the United States show that 44,000 ft was the highest altitude the hail was encountered. More encounters occurred at 6,000 to 10,000 ft. More than 40% of encounters occurred above about 20,000 ft and 16% above 30,000 ft. This suggests that there is preponderance of large hailstones when hail is formed in the upper troposphere. Adequate information is not available about hail size at flight heights because pilots usually avoid the thunderstorm zone in the troposphere. But Foster (1961) reported within his limitations that 5-in hail is formed up to 29,500 ft and 4-in hail is formed at 31,000 ft and 3-in hail is formed at 37,000 ft. It is reported by Donaldson (1959, 1961) that the larger the vertical extent of thunderstorm, the probability of getting hails at the ground is large. The size of the hails available at the ground is summarized in Tables 7.5 and 7.6. The probability of occurrence of hail over India is also summarized in Table 7.6. It is to be remembered that with

TABLE 7.5

Sizes of Hail

Place	Size
India	0.2–1.2 inches
Central Europe	0.1–1.2 inches
USA	0.5–0.7 inches

TABLE 7.6

Size Distribution of Hail

Less than 0.2 inches	27%
0.2–1.2 inches	51%
Greater than 1.2 inches	22%

the increasing intensity, hail can significantly degrade millimeter wave propagation. Limited information is available in this regard. Although limited, the information provided about hail can be utilized for getting at least first-hand approximation about attenuation.

7.3 Fog

Fog can be formed in variety of ways depending mostly on the condensation mechanism. Generally, we encounter the type of fog that is formed by the cooling of land after sunset by thermal radiation in calm and clear sky. The cool ground produces condensation in the nearby area by conduction of heat. This type of fog mostly prevails at night and does not last long after sunrise. It generally occurs in autumn and winter.

Fog attenuation is comparatively smaller at millimeter wavelength. It is composed of suspended spherical water droplets with radii small enough ($r < 50$ μm) to keep them suspended in air by micro turbulence (Liebe et al. 1989). Attenuation due to fog in the millimeter wave band is mainly caused by absorption and scattering, which in turn depends on the extent of the fog and its index of refraction. According to Gibbins (1988), the fog density is given in terms of visibility V (km) by

$$M = \left(\frac{0.024}{V} \right)^{1.54} gm^{-3} \qquad (7.4)$$

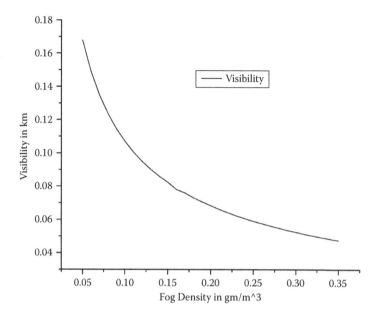

FIGURE 7.3
The variability of visibility in presence of fog.

However, let us confine ourselves with the radiation fog for which the fog density approaches nearly 1gm/m³. But over the United Kingdom, Gibbins (1988) pointed out that fog occurs typically of the order of 1% to 2% of the time over a period of one year. The density of this type of fog is 0.02 gm/m³ for which the visibility is 300 m and is termed as moderate fog. The variation of extent of fog in terms of density is shown in Figure 7.3.

Comparing the size of the suspended water droplets (fog) and the wavelength in the millimeter wave band at 94 GHz (conventional window frequency), the Rayleigh approximation may be adopted for defining the refractivity model. Using the Rayleigh absorption approximation (Van de Hulst 1957), refractivity is written as

$$N = \frac{M\left(\frac{3}{2m_w}\right)[\varepsilon - 1]}{[\varepsilon - 2]} ppm \tag{7.5}$$

where m_w is the specific weight for water (i.e., 1.0).

The complex permittivity is

$$\varepsilon = \varepsilon'(f) - j\varepsilon''(f) \tag{7.6}$$

The refractivity term N is subdivided into three parts among which the nondispersive refractivity is given by (Liebe et al. 1989)

$$N_0 = M\left(\frac{3}{2}[1 - 3/(\varepsilon_0 + 2)]\right)$$

$$N'(f) = M\left(\frac{9}{2}\right)\left[\frac{1}{(\varepsilon_0 + 2)} - y/\varepsilon''(y^2 + 1)\right]$$

And the loss term is given by

$$N''(f) = M\left(\frac{9}{2}\right) \times \frac{1}{[\varepsilon''(y^2 + 1)]} \tag{7.7}$$

where
$$y = (\varepsilon' + 2)/\varepsilon''$$
$$\varepsilon_0(T) = 77.66 + 103.3(\theta - 1)$$
$$\theta = 300/T(K)$$

The permittivity of liquid water is

$$\varepsilon = (\varepsilon_0 - \varepsilon_1) \times \left[1 + \left(\frac{f}{f_p}\right)\right] + \frac{(\varepsilon_1 - \varepsilon_2)}{\left[1 + j\left(\frac{f}{f_s}\right)\right]} + \varepsilon_2 \tag{7.8}$$

Real and imaginary parts of Equation (7.8) are

$$\varepsilon'(f) = \frac{(\varepsilon_0 - \varepsilon_1)}{\left[1 + \left(\frac{f}{f_p}\right)^2\right]} + \frac{(\varepsilon_1 - \varepsilon_2)}{\left[1 + \left(\frac{f}{f_s}\right)^2\right]} + \varepsilon_2 \tag{7.9}$$

and

$$\varepsilon''(f) = \frac{(\varepsilon_0 - \varepsilon_1)\left(\frac{f}{f_p}\right)}{\left[1 + \left(\frac{f}{f_p}\right)^2\right]} + \frac{(\varepsilon_1 - \varepsilon_2)\left(\frac{f}{f_p}\right)}{\left[1 + \left(\frac{f}{f_s}\right)^2\right]} \tag{7.10}$$

where f_p = principal relaxation frequency

$$= 20.09 - 142.4(\theta - 1) + 294(\theta - 1)^2 \text{ GHz} \tag{7.11}$$

and $f_s(T)$ = secondary relaxation frequency

$$= 590 - 1500\ (\theta - 1) \tag{7.12}$$

The other two permittivity constants are $\varepsilon_1 = 5.48$ and $\varepsilon_2 = 3.51$.

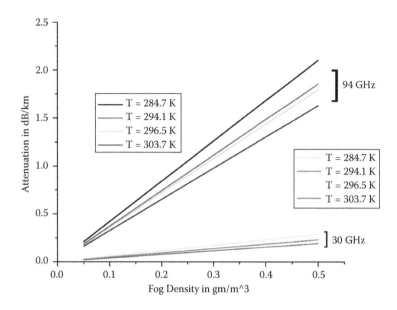

FIGURE 7.4
Fog attenuation at 30 and 94 GHz at temperatures ranging from 284 to 303 K.

Now using the idea of refractivity as a fundamental parameter to practical propagation terms such as power loss we write (Liebe et al. 1989)

$$\alpha = \frac{0.182\, f\, N''db}{km} \qquad (7.13)$$

where N'' is described in Equation (7.6).

Fog attenuations at 30 and 94 GHz are presented in Figure 7.4, considering temperatures as the parameter ranging from 284 K to 303 K. It is clear from the figure that at all the said temperature regions the 30 GHz attenuation never exceeds 0.35 dB/km (Karmaker et al. 2010). But on the other hand, the 94 GHz attenuation shows that, for the same range of temperatures and for fog densities 0.5 gm/m^3, its value ranges from 2.1029 to 1.6283 dB/km. For the sake of clarity, it is also observed that for 0.5 gm/m^3 and for temperature 284.7 K the ratio of 94 GHz attenuation is more or less 7 times larger than that at 30 GHz. This ratio becomes very much pertinent to our study because we are concerned with the radiation fog whose density mostly approaches toward 0.3–0.7 gm/m^3. But for the advection fog whose density is less (nearly equal to 0.2 gm/m^3), this attenuation ratio is not prominent enough to be noticed.

The dependence of temperature on fog attenuation (Figure 7.5) is presented keeping fog density equal to 0.5 gm/m^3.

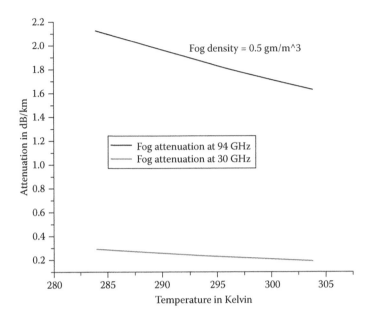

FIGURE 7.5
Variation of fog attenuation at 94 and 30 GHz with temperature keeping fog density 0.5 gm⁻³.

7.4 Aerosols

Aerosols may be defined as a mixture of air and small solid or liquid particles in which the particle will follow the motion of air. Visible haze is the evidence of aerosol content. Fog and cloud sometimes may be called aerosols but they remain on the border line depending on their size. Rain and snow are not the aerosols because they can be separated rather than suspension in the air. Basically, the products of air pollution through photochemical reaction generate aerosols. Atmospheric aerosols affect the earth's energy balance primarily through scattering and absorption of short-wave radiation and by acting as seed particles in cloud-forming processes (Tzanis and Varotsos 2008). They remain mostly in the size range 10^{-3} to 10^2 microns. In fact, detailed and quantitative knowledge of earth's radiation field is crucial to understand and predict the evolution of the components of earth's system (Varotsos et al. 1995, 2000, 2003; Ricchiazzi et al. 1998; Varotsos 2005). The propagation of radio waves through aerosols is very complex. Care has to be taken because of local aerosol characteristics. Generally, populations are made up of individual units with diverse characteristics. Atmospheric aerosol size distribution, composition, morphology and source strengths as well as their vertical distribution can vary with meteorology, location, and time (Pekney et al. 2006). Due to the lack of information of their global distribution and detailed knowledge of their optical properties, aerosols are considered as a major uncertainty in climate forcing assessments

(Charlson et al. 1992; Hansen et al. 2000). However, the atmospheric aerosols not only scatter and absorb solar radiation but also absorb and emit long-wave radiation. This, in turn, alters the atmospheric heating rate and hence changes the atmospheric circulation (Alpert et al. 1998; Miller and Tegen 1998). For example, the thermal infrared radiative effects of dust are about 10% of those for short-wave radiation (Tanre et al. 2003) and the long-wave aerosol forcing is estimated to be only 0.01 Wm^{-2} on a mean global basis (Haywood et al. 1997). It is also found by Hansen et al. (1997) that in cloudy conditions the aerosol radiative effect depends on the fraction of absorbing aerosols located above clouds where the particles can absorb up to three times more sun radiation than in a cloud-free condition. The direct effect of aerosol on short-wave radiation in the presence of clouds is smaller than in clear sky (Haywood and Ramaswamy 1998) except for soot aerosols above clouds and sulfate aerosols below cirrus clouds (Highwood 2000). The model population available in this regard can only be used as a guideline because variation of an order of magnitude or more can influence the propagation of radio wave over a particular place of choice. But, in any case we need to have a size distribution of aerosols to estimate the effect of aerosols on propagation of radio waves. Assuming the spherical aerosols, we consider n is the total number of particles per cc with radii from lower limit, defined arbitrarily up to a radius r. The logarithmic derivative $dn/d(\log r)$ is called the size distribution. The size distribution versus $\log r$ is presented in Figure 7.6. Because of this log-log curve, the area under the curve should not represent the number of particles chosen between the corresponding radii. To get the number distribution or size distribution between two values r_1 and r_2, the following procedure is adopted, where,

$$\Delta n\big|_{r_1}^{r_2} = \frac{dn}{d(\log r)} \log(r_2 - r_1) \tag{7.14}$$

The portion of the distribution curve between $-1 < \log r < + 1$ (particle radius 0.1 to 10 micron) can be represented as

$$\frac{dn}{d(\log r)} = Cr^{-3} \tag{7.15}$$

On integration, we get

$$\Delta n\big|_{r_1}^{r_2} = \frac{1}{3\ln 10} \left[\frac{dn}{d(\log r)} \right]_{r_1}^{r_2} \tag{7.16}$$

To exemplify, we consider two values of r and let it be $r_1 = 0.1$ microns and $r_2 = 0.2$ microns. Then from Equation (7.16), we write

$$\Delta n| = \frac{1}{6.9}(4000 - 600)$$

$$= 490 \, cm^{-3}$$

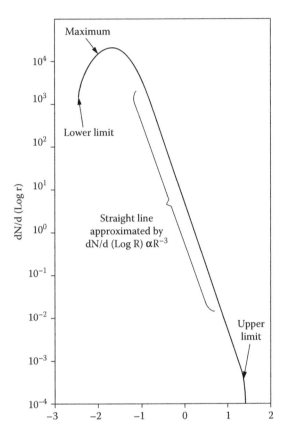

FIGURE 7.6
Variation of aerosols size distribution with those of log (*r*).

Hence, by using this size distribution of aerosols we can calculate the required attenuation of radio waves as we did earlier in case of attenuation with other precipitation particles. However, this attenuation depends on aerosol optical depth and in this regard, Tzanis and Varotsos (2008) performed an observation of average optical depth over Crete, Greece (Eastern Mediterranean) for the period 2000–2001 using Total Ozone Mapping Spectrometer (TOMS), which was on board the Nimbus-7 Meteor-3 Earth Probe and Advanced Earth Observing Satellite. They found that the aerosol detection capability in the near UV first became apparent with the development of the TOMS aerosol index as a by-product of TOMS version 7 ozone algorithm. This index is a measure of change of spectral contrast in the near UV due to radiative transfer effects of aerosols in a Rayleigh scattering atmosphere (Torres et al. 2002). On the other hand, the aerosol optical depth and surface albedo data over greater Athens (35.59° N) were incorporated in a discrete ordinate radiative transfer mode developed by University of Santa Barbara (Ricchiazzi

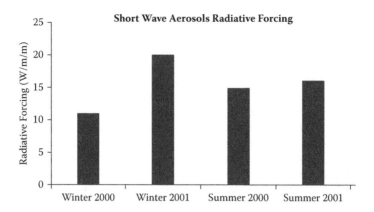

FIGURE 7.7
Short-wave radiative forcing at the surface (W m⁻²) due to aerosols for 2000–2001 over Athens, Greece.

et al. 1998) to estimate the aerosol impact on the short-wave and long-wave fluxes under clear sky conditions. Figure 7.7 shows a large reduction in surface reaching solar radiation as much as 11–20 Wm⁻² due to both natural and anthropogenic aerosols over Greece (Tzanis and Varotsos 2008). The maximum of short-wave aerosol forcing in winter 2001 is due to high values of aerosol optical depth, about two times larger than that with the optical depth in winter 2000. Tzanis and Varotsos (2008) have presented the vertical profile of the radiative heating rates due to aerosols for winter and summer months over Greece (Figure 7.8).

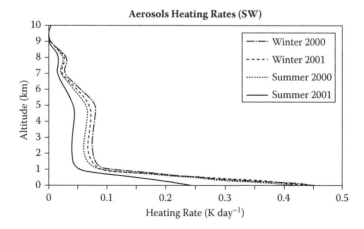

FIGURE 7.8
Radiative heating rates due to aerosols for different altitudes in the troposphere over Athens, Greece.

7.5 Clouds: Nonprecipitable Liquid Water

We are usually accustomed with different types of clouds discussed by several authors along with their generation or formation mechanism. But the thunderstorm cloud affects the radio wave in a major way. This type of cloud sometimes extends well above the cirrus type and penetrates a few thousand feet into the stratosphere. These clouds must contain exceedingly high vertical velocities and in that case the presence of hail above the tropopause is also possible. Sometimes it also happens that under steady conditions the mass of cloud water per unit volume may be two or three times than that of rain in the zone just below the melting level, especially in case of light precipitation. This factor should be counted for in radio wave propagation. The water content of cloud normally increases upward to a maximum in the vicinity of the melting level and then above a temperature level of one degree or two degree centigrade and gradually decreases to zero.

But it is to be pointed out that such steady-state conditions with respect to water distributions in the cloud do not exist when updrafts equal or exceed the fall velocity of particle as rain. Such updrafts exist locally for periods about 5 to 15 minutes in thunderstorm or other convective activity and hence can lead to high local concentration of water. Radar reflectivity data as shown by Donaldson (1961) shows peak water profiles during a thunderstorm over New England (Figure 7.9) by using the approximation

$$M = (6 \times 10^{-4})Z^{0.7} \tag{7.17}$$

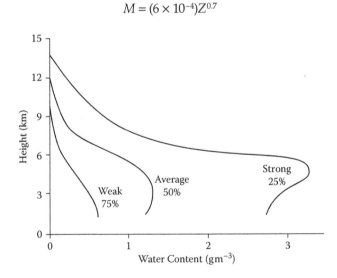

FIGURE 7.9
Profiles of concentrations of water in the centers of mild, average, and strongly reflective New England thunderstorm. Figures on the curves refer to fraction of the thunderstorms with water content profiles in excess of snow.

where Z is the radar reflectivity and M is the liquid water content. It is clear also from Figure 7.9 that the occasional severe thunderstorm may be associated with local water concentrations, which is about ten times larger than the highest values shown in Figure 7.9.

7.5.1 A General Idea of Maximum Water Content in Clouds

Generally, water in clouds exists in vapor, liquid and solid ice crystal states. Amongst these, vapor exists at temperatures. Liquid water is found in clouds from 25°C down to −35°C or even to −40°C. Ice crystals are found at all sub zero temperatures but generally will not form in the free atmosphere at temperatures warmer than −12°C.

Existence of water vapor in the atmosphere always (for practical purposes) is indicated by humidity. Relative humidity of clouds is 100%. The amount of water vapor depends on cloud temperature, doubling or tripling for each 10°C increase in temperature. If clouds at 15°C contain 23 gm–m^{-3} of vapor, then at 0°C, its value may come down to 5 gm–m^{-3} only.

As the amount of vapor approximately doubles, for each 10°C rise in temperature, more water will be available during summer and heavier clouds are to be expected below 2500 ft.

Aufm Kampe and Weickmann (1957) show that warm convective cloud contains an average liquid water 4–5 times to that observed in winter and 5–10 times to that observed in stratus cloud irrespective of season. It was also observed by the same author that maximum water content over New Jersey and Florida during summer is 10 gm–m^{-3} at temperature 108°C–8°C in cumulonimbus cloud at about 4000 m above cloud base. But in case of cumulus congestus cloud, it obtains a maximum of 6.6 gm–m^{-3}. A report from University of Chicago indicates that cloud water densities of at least 1.7 gm–m^{-3} are required before rain is produced.

7.6 Microwave Radiometric Estimation of Water Vapor and Cloud Liquid

Water vapor is perhaps the most important minor constituent that can affect the thermodynamic balance, photochemistry of the atmosphere, sun–weather relationship, and the biosphere as well. Measurements of the vertical and horizontal distribution of water vapor as well as its temporal variation are essential for probing into the mysteries of several effects. The conventional method of measuring water vapor by using radiosonde data is not continuous and of very limited accuracy. In this context, the ground-based microwave radiometric sensing may provide the continuous and true values of ambient water vapor.

It is a common practice to undertake the single frequency measurement at the peak of the microwave absorption spectra. The first absorption spectra occur at 22.235 GHz. So one can have the choice of exploiting 22.235 GHz on the basis of assumption that the signal to noise ratio is largest at this frequency provided pressure and temperature profiles are constant (Resch 1983). But this does not happen in practice. Pressure and temperature are highly variable parameters of the atmosphere. Wesrtwater (1980) showed by using the Van Vleck line shape and by using Liebe's parameter and again for Zhevaking-Naumov gross line shape with the parameter given by Waters, that the frequency independent of pressure lie both ways around the resonance line, that is, 22.234 GHz. Moreover, this single frequency measurement of water vapor will be influenced by the presence of overhead cloud liquid.

It has been shown (Simpson 2002) that a zenith-pointing ground-based microwave radiometer measuring sky brightness temperature in the region of 22 GHz is three times more sensitive to the amount of water vapor than the amount of liquid water. However, in the region of 30 GHz the sky brightness temperature is two times more sensitive to liquid water than that of water vapor, with the consideration that the sensitivity to ice is negligible at both frequencies.

Measurements of the vertical and horizontal distribution of water vapor, as well as its temporal variation are essential for probing into the mysteries of several atmospheric effects. A sizeable literature has focused on the complex relationship between water vapor variability and deep convection in the tropics (Sherwood et al. 2009). Unlike higher latitudes, rotational dynamical constraints are weak and precipitation-induced heating perturbations are rapidly communicated over great distances. Water vapor, on the other hand, is highly variable in space and time; its spatial distribution depending on much slower advection processes above the boundary layer and on deep convection itself. Furthermore, deep convection, through vertical transport of water vapor and evaporation of cloud droplets and hydrometeors, serves as the free tropospheric water vapor source. And deep convection is itself sensitive to the free tropospheric humidity distribution through local moistening of the environment that favors further deep convection; a positive feedback (Adams et al. 2011). In this context, the ground based microwave radiometric sensing appears to be one of the suitable solutions for continuous monitoring of ambient atmospheric water vapor.

7.6.1 Single Frequency Algorithm for Water Vapor Estimation

The vertical and horizontal distribution of water vapor will vary in both space and time. Resch (1983) calculated the atmospheric emission around 22 GHz for two different vertical distributions of water vapor assuming the pressure distribution in a standard atmosphere and standard lapse rate. The brightness temperature was calculated for a constant relative humidity of 81.6% and for the height region 0–1 km (Figure 7.10a); and for 99%

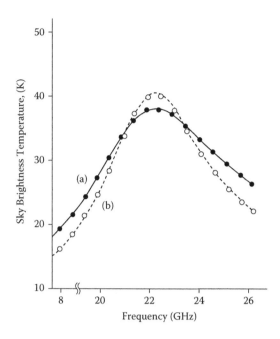

FIGURE 7.10

Line profiles of atmospheric water vapor for two different vertical distributions in standard atmosphere. (a) $RH = 81.6\%$ for $0 < H < 1000$ m (filled circles). (b) $RH = 99\%$ for $1000 < H < 3800$ m (open circles).

humidity and for height region 1.0–3.8 km (Figure 7.10b). The high altitude vapor clearly shows a sharper line than the low altitude profile, although the delays in both cases were the same. Hence it may be suggested that if one wishes to maximize the signal from a given amount of water vapor then clearly the observation should be carried at 22.235 GHz. It is also clear from the figure that a single frequency measurement of brightness temperature near the half power point of the line profile would provide the most accurate estimates. But, it is to be remembered that the following algorithm should be valid only in cases of a cloudless and clear sky.

The absorption (cm^{-1}) of a water vapor molecule at 22.235 GHz is given by Bhattacharya (1985) as

$$\Gamma = 3.24 \times 10^{-4} \frac{P \exp\left[-\frac{644}{T}\right] \gamma^2}{T^{3.125}} \times \left[1 + 0.0147 \frac{\rho T}{P}\right]$$

$$\times \left[\frac{1}{(\gamma - 22.235)^2 + (\Delta \gamma)^2} + \frac{1}{(\gamma + 22.235)^2 + (\Delta \gamma)^2}\right] \qquad (7.18)$$

$$+ 2.55 \times 10^{-8} \frac{\rho \gamma^2 \Delta \gamma}{T^{3/2}}$$

where γ is the frequency in GHz, T is the kinetic temperature in K, and $\Delta\gamma$ is the pressure broadened line half width parameter and is given by

$$\Delta\gamma = 2.58 \times 10^{-3} \left[1 + 0.0147 \frac{\rho T}{P} \right] \frac{P}{\left(\frac{T}{318}\right)^{0.625}} \text{ GHz} \qquad (7.19)$$

where P is the total pressure in mm of Hg and ρ is the water vapor density (gm^{-3}).

After simplification, Equations (7.18) and (7.19) yield

$$\Gamma = 17.92 \frac{\rho \exp\left[\frac{-644}{T}\right]}{PT^{1.875}} \times \left[1 + 0.0147 \frac{\rho T}{P} \right]^{-1}$$

$$+ 11.91 \times 10^{-7} \left[1 + 0.0147 \frac{\rho T}{P} \right] \frac{\rho}{T^{2.125}} \ cm^{-1} \qquad (7.20)$$

In Equation (7.20), the first term is the resonance term and the second term is the nonresonance term. Considering the typical surface parameters, $P = 750$ mm of Hg; $T_0 = 300$ K; and $\rho_0 = 25$ gm^{-3}, we find the contribution of the nonresonance part is of the order of 1% of the resonance part. Hence, the nonresonance part is neglected in comparison to the resonance part. Thus, we are left with

$$\lambda = 17.92 \frac{\rho \exp\left[\frac{-644}{T}\right]}{PT^{1.875}} \times \left[1 + 0.0147 \frac{\rho T}{P} \right]^{-1} \ cm^{-1} \qquad (7.21)$$

Moreover, using the same data, we find that the value of $0.0147 \ \rho T/P = 0.145$ is much less in comparison to the first term and hence is neglected. Now, converting Equation (7.21) into dBkm^{-1}, we get

$$\lambda = 17.92 \left[\log_{10} \rho \times 10^6 \right] \times \frac{\rho \exp\left[\frac{-644}{T}\right]}{PT^{1.875}}$$

$$= 7.78 * 10^6 \times \frac{\rho \exp\left[\frac{-644}{T}\right]}{PT^{1.875}} \qquad (7.22)$$

$$= 7.78 \times 10^6 \times \rho(T) \times F(T) \times \frac{T^{0.52699}}{P} \ dBkm^{-1}$$

where $F(T) = \frac{\exp[-644/T]}{2.40199}$ is an implicit function of temperature (Karmakar 1989).

The range of temperature (T) over Kolkata is such that $F(T)$ represents a slowly varying function, as depicted in Figure (7.11). It is found there that the

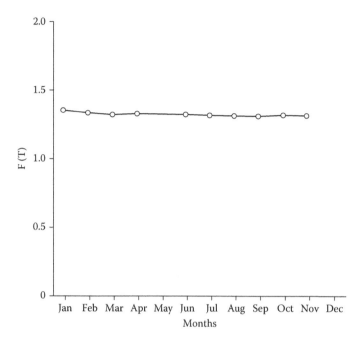

FIGURE 7.11
Monthly variation of $F(T)$. It is an implicit function of atmospheric temperature, T.

monthly variation of T makes $F(T)$ a slowly varying function over the range of temperature which is considered to be the typical variation of the parameter over Kolkata during the full calendar year. Now, referring to Figure (7.11), we have taken the liberty to take the value of $F(T)$ equal to 1.3×10^{-7} (average value). Hence, Equation (7.22) reduces to

$$\lambda = 7.78 \times 10^6 \times 1.3 \times 10^{-7} \times \frac{\rho T^{0.52699}}{P} \, dBkm^{-1}$$

$$= 1.0114 \times \frac{\rho T^{0.52699}}{P} \, dBkm^{-1}$$

On integration, we get

$$A = 1.0114 \int_0^\alpha \frac{\rho T^{0.52699}}{P} \, dh (dB) \qquad (7.23)$$

Here A represents the total atmospheric water vapor attenuation in dB along a vertical path. To carry out the integration, according to Hess (1959) and Braunt (1947), the atmosphere is assumed to be of constant lapse rate. In an atmosphere having a constant lapse rate the relation between temperature

(T) and pressure (P) is given by Poisson's equation, which may approximately be written as

$$T = T_0 \left[\frac{P}{P_0} \right]^{R\beta/g} \tag{7.24}$$

$$P = P_0 \exp\left(-h / H_p \right) \tag{7.25}$$

$$\rho = \rho_0 \exp\left(-\frac{h}{H_\rho} \right) \tag{7.26}$$

where R is the gas constant; β is the lapse rate; T_0, P_0, and ρ_0 are the surface parameters; H_p and H_ρ are the pressure scale height and water vapor scale height, respectively; and g is the acceleration due to gravity. From Equation (7.23)

$$A = 1.0114 \times \frac{\rho_0 T_0^{0.52699}}{P_0} \int_0^\alpha \exp(-H_1 h)\, dh$$

$$= 1.0114 \times \frac{\rho_0 T_0^{0.52699}}{P_0 H_1}$$

where

$$H_1 = \frac{1}{H_\rho} - \frac{1 - (0.52699 R\beta/g)}{H_\rho} \tag{7.27}$$

For Kolkata, $\beta = 0.7509\text{K}/100$ m, average adiabatic lapse rate for 1.5 km to 7 km (data taken from Civil Aviation Department, Kolkata Airport; $R = 2.9$ units and, $H_p = 8$ km).

From Equations (7.25) and (7.26) and substituting R, H_p and β we get

$$\frac{P_0 A}{H_\rho} = 1.0114 \times T_0^{0.52699} \times \rho_0 + P_0 \times 0.1103 A$$

$$H_\rho = \frac{P_0 A}{1.0114 \times T_0^{0.52699} \times \rho_0 + P_0 \times 0.1103 A} \text{ km} \tag{7.28}$$

where A is the zenith attenuation (dB) to be measured by using the 22.235 GHz radiometric data.

However, according to Allnutt (1976), the zenith attenuations may be calculated from the radiometric output in the form of brightness temperature by using the equation

$$A = 10 \log_{10} \frac{T_m - T_{cos}}{T_m - T_a(f)} \tag{7.29}$$

where $T_a(f)$ represents the brightness temperature at a frequency of choice. Here, T_m is the mean atmospheric temperature, which is eventually found to be dependent on ground temperature, obeying the relationship $T_m = C(f)T_s$. Here C is considered to be a frequency-dependent term. Mitra et al. (2000) found $C(22.235) = 0.95$. So with the available data for T_s (surface temperature), the values of T_m may be calculated for different occasions. Now with the accepted values of, $T_{cos} = 2.75$ K (noise temperature due to cosmic background), the attenuation values from the measured radiometric brightness temperature T_a (refer to Equation 7.28) may be calculated. Hence by using Equation (7.28), the appropriate values of the water vapor scale height may be determined.

The amount of water vapor contained in the atmosphere is a function of several meteorological parameters, but it specially depends on the atmospheric temperature. The primary parameter of interest in finding the water vapor density is the partial water vapor pressure (mb), which is given by the relation (Moran and Rosen 1981)

$$e = 6.10 \exp\left\{ 25.228\left(1 - \left(\frac{273}{T_D}\right)\right) - \left(5.31 \log\left(\frac{T_D}{273}\right)\right)\right\} \tag{7.30}$$

where T_D is the dew point in K. The attenuation rate due to the water vapor monomer model in the atmosphere at 22.235 GHz may be estimated on the basis of Equations 7.20 and 7.21. The oxygen contribution to the radiometric noise is dependent only on the pressure and temperature, and would be of a very slowly varying nature in comparison to water vapor contribution. According to Buttorn and Wiltse (1981), the oxygen contribution approximately is 0.06 dB. So this would be the additive term with the vapor attenuation term for which the contribution to brightness temperature is only about 2–3 K.

Karmakar et al. (1989) found that the height distribution of the meteorological parameters are obeying the following empirical relationships:

$$e(mb) = e_0 \exp(-m_1 h)$$
$$\rho(g - m^{-3}) = \rho_0 \exp(-m_2 h)$$
$$P(mb) = P_0 \exp(-m_3 h) \tag{7.31}$$
$$T(K) = T_0 - m_4 h$$

Here m's are the scale factors.

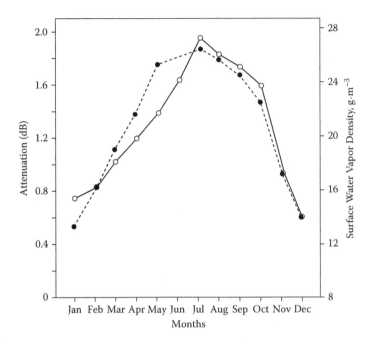

FIGURE 7.12

Monthly variation of calculated attenuation (dB) and corresponding monthly variation of surface water vapor density (gm⁻³) over Kolkata (22° N).

The monthly variation of calculated attenuation (dB) and the corresponding monthly variation of surface water vapor density (gm⁻³) over Kolkata (22° N), India, are shown in Figure 7.12. The monthly variation of calculated and measured brightness temperature at 22.235 GHz over Kolkata is shown in Figure 7.13. Both figures show that the brightness temperatures and water vapor density bear a maximum during the month of July through August over Kolkata. This is presumably due to maximum abundances of water vapor in these months. A comprehensive study has been made to compare the calculated values of attenuations obtained by different authors (Waters 1976; Bhattacharya 1985; Liebe 1989). The results are summarized in Table 7.7. If we look back to the Figures 7.12 and 7.13, we will see some discrepancies between measured and calculated values. This might need the inclusion of an empirical, nonresonant correction, which depends on the square of water vapor density. But still, the origin of this effect is not properly understood. The possible reason has been sought in terms of hydrogen bonding of the water molecule to form a dimer (Bohlander et al. 1980) or of the water molecules together or in terms of errors in the line shape used in the calculation (Gibbins 1986).

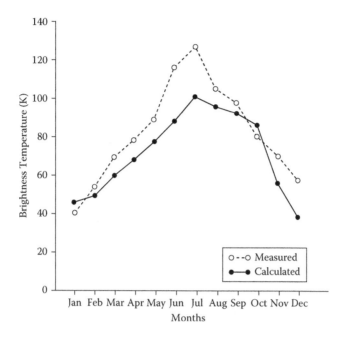

FIGURE 7.13
Monthly variation of calculated and measured brightness temperature, for clear sky only.

7.6.2 Water Vapor Scale Height

Analytically, the water vapor scale height is defined as the height at which water vapor density becomes $1/e$ times the surface value, that is, ρ_0 (refer to Equation 7.21). The monthly variation of water vapor scale height has been shown by Sen et al. (1990) over Kolkata (Figure 7.14) and that over the National Institute for Space Research (INPE) in Brazil (22° S; Figure 7.15) by Karmakar et al. (2010). Figure 7.14 shows that the average values of calculated

TABLE 7.7

Comparative Studies of Attenuation over Kolkata Using Different Models

Month	Liebe (1989)	Waters (1976)	Bhattacharya (1985)
January	0.65339	0.665	0.6143
March	1.065	1.072	1.035
May	1.391	1.390	1.386
June	1.724	1.703	1.716
July	2.036	2.027	2.0235
September	1.7084	1.676	1.683
November	0.9403	0.9533	0.905

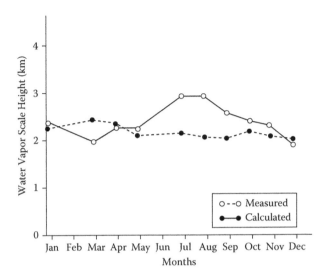

FIGURE 7.14
Monthly variation of calculated and measured water vapor scale height over Kolkata (22° N).

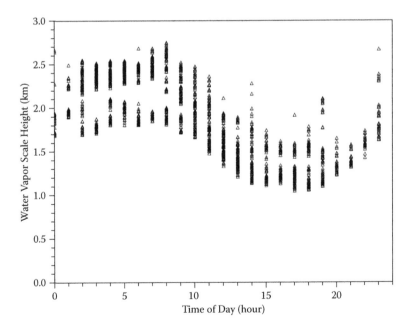

FIGURE 7.15
A time series of water vapor scale height on April 15 over Brazil (22° S) during clear sky condition.

and measured water vapor scale height are 2.45 km and 2.24 km. This also shows that during May through October, the calculated value overestimates the measured ones. This anomaly might be due to inconsistent radiosonde data. However, it may be decided that the water vapor scale height varies over Kolkata, India, from 2.0 to 2.5 km. But care has to be taken about the rainy season when the algorithm developed so far is not applicable for the measurement of water vapor. It is also clear from Figure 7.15 that the water vapor scale height values over Brazil attain a minimum around 17 hours of the day with no overhead cloud. However, the presence of cloud may produce a deviation in this pattern. It may also happen that the suspended, nonprecipitable liquid water particle (density <1 g/c.c.) to be evaporated by taking necessary latent heat from the atmosphere. So if we can measure the water vapor scale height from radiometric measurement (refer to Equation 7.28), it would then be convenient to get the vertical profile of water vapor. Hence, by integrating the water vapor profile within the height limit 0–10 km, one can get the integrated water vapor content where we define the mass of water vapor content in a cylindrical column of one square meter cross-sectional base area as integrated precipitable water vapor expressed in kg/m².

7.6.3 Dual Frequency Algorithm for Water Vapor and Cloud Liquid Estimation

Signal attenuation in the microwave region is mainly due to water vapor, liquid water, and gaseous particles (mainly oxygen). Among these, the water vapor plays a major role in attenuating the microwave signal during its propagation through the ambient atmosphere. The microwave signal also gets attenuated in the presence of clouds containing liquid water. Gaseous oxygen, though not as intense as water, contributes to the attenuation. Now, the total attenuation can be written mathematically as (Karmakar et al. 2001)

$$A_T = A_L + A_O + A_V \qquad (7.32)$$

where,

A_T = Total attenuation (dB)
A_L = Attenuation due to nonprecipitable cloud liquid water
A_O = Attenuation due to oxygen
A_V = Attenuation due to water vapor

Now we write, using Equation (7.32), total attenuation at f_2 GHz (higher frequency) as

$$A_T(f_2) = A_L(f_2) + A_O(f_2) + A_V(f_2) \qquad (7.33)$$

And the total attenuation involving at f_1 GHz (lower frequency) is

$$A_T(f_1) = A_L(f_1) + A_O(f_1) + A_V(f_1) \qquad (7.34)$$

Attenuation due to cloud liquid at the two frequencies are assumed to be related as

$$A_L (f_2) = K_1 \times A_L (f_1) \tag{7.35}$$

where $K_1 = (f_2/f_1)^2$

Multiplying Equation (7.34) by, K_1 we get,

$$K_1 A_T (f_1) = K_1 A_L (f_1) + K_1 A_O (f_1) + K_1 A_V (f_1) \tag{7.36}$$

From Equations (7.33) and (7.36), and rearranging we get

$$[K_1 A_T (f_1) - A_T(f_2)] - [K_1 A_O (f_1) - A_O (f_2)] = K_1 A_V (f_1) - A_V (f_2) \tag{7.37}$$

Now we choose the left-hand side of the Equation (7.37) to be written as

$$Z_v = [K_1 A_T (f_1) - K_1 (f_2)] - [K_1 A_O (f_1) - A_O (f_2)] \tag{7.38}$$

Here we see that the term Z_v contains only the vapor part and dry part, and can be termed as attenuation part free from liquid attenuation. A statistical least square fitting is adopted between the calculated values of V obtained by using radiosonde data (refer to Equation 7.46) and the measured values of attenuations at desired frequencies (refer to Equation 7.38) and we get

$$V(kgm^{-2}) = Z_v p_v + q_v \tag{7.39}$$

Here it is to be mentioned that the oxygen contribution has to be taken care of as was done in the previous section.

A statistical regression analysis between the calculated values of water vapor attenuation $A_V (f_1)$ and $A_V (f_2)$ by using radiosonde data reveals that they are related as

$$A_v f_2 = K_2 A_v (f_1) \tag{7.40}$$

Again, involving Equations (7.40) and (7.41), and progressing similar to the case of water vapor we get

$$Z_L = [A_T (f_2) - \{k_2 A_T (f_1)\}] - [A_O (f_2) - \{k_2 A_O (f_1)\}] \tag{7.41}$$

The liquid water content L is obtained from the Equation (7.51) (refer to the next section). This L and Z_L are related by a linear relationship

$$L = Z_L p_L + q_L \tag{7.42}$$

where p_L and q_L are regression constants.

7.6.3.1 Use of Radiosonde Data

Liquid water is the depth of water that could be collected from a column of cloud liquid water droplets, which can also be expressed in kg/m². This measurement cannot be done analytically. So it is accomplished by constructing a database of water vapor content, liquid water content, and the brightness temperature values. Then derivation of the necessary coefficients may be performed using multiple linear regressions. Measurements of brightness temperature from the radiometers can then be transformed into estimates of ambient water vapor content and liquid water content.

The vertical profiles of temperature $T(h)$ in Kelvin, pressure $P(h)$ in Pascal, and dew point temperature T_d in Kelvin may be obtained from the radiosonde data. The partial water vapor pressure e can be determined by using the Equation 7.30.

But during saturation, this water vapor pressure can be formulated as

$$e_s = 6.112 \frac{\exp(17.502 \times t)}{t + 240.97} \quad mb \tag{7.43}$$

where t is the atmospheric temperature in degree centigrade.

Now, one can take the liberty to consider that water vapor approximately behaves as an ideal gas, where each mole of gas obeys an equation of state that can be written as

$$\rho_v = \frac{e}{R_w} T \tag{7.44}$$

where ρ_v is the water vapor density (kg/m³) and e is the partial pressure of water vapor (mb). R_w is the gas constant for water vapor, that is, $R_w = R/m_w$, where R is the universal gas constant and m_w is the mass of one mole of vapor. T is the absolute temperature (K). Here $R = 8.135$ J/mole K and $m_w = 18$ gm. Substituting all these values in Equation (7.43), we find the water vapor density (gm⁻³)

$$\rho_V = \frac{1800e}{8.3145T} \tag{7.45}$$

Usually ρ_V is multiplied by 10^3 and expressed in kg/m³. However, relative humidity (RH) at different height levels at the place of interest were found by using the relation

$$RH = \frac{e}{e_s} \times 100\%$$

In the present context, we are interested in finding the integrated water vapor content, defined earlier, that can be written as

$$V(\text{kg/m}^2) = \int_0^\infty \rho_v(h)\,dh \qquad (7.46)$$

Water vapor absorption coefficients (α_V dB/km) at the corresponding heights may be obtained using Liebe's MPM model (Liebe 1985). The data inputs are temperature, pressure, and *RH*.

Each radiosonde ascent records profiles of atmospheric pressure, altitude, temperature, and dew point temperature data. In the standard data format used, all four quantities are generally recorded at 15 specified values of atmospheric pressure. Intermediate values may be recorded whenever there are significant changes of pressure or temperature at corresponding altitudes. It can be determined by interpolation, assuming an exponential profile between the relevant pressure values. Vertical profiles that match the slab heights required by the Liebe model are then obtained by further interpolation. It is to be noted that the atmosphere may be divided into 0.2 km thick slabs and above 5 km it may be assumed to be 1 km thick. The pressure, temperature, and dew point temperature at each height are used to derive relative humidity, air density, and vapor pressure for each slab. From these we may obtain the water vapor density profile, and hence vapor content by integration with respect to height.

Brightness temperature (T_b) at the desired frequency pair may be obtained by using Equation (7.29) (Ulaby et al. 1986). The mean atmosphere temperature (T_m), a frequency dependent term, may be found by using the relation

$$T_m = \frac{\int T(z)\alpha(z)\exp\left\{-\int \alpha(z)dz\right\}dz}{\int \alpha(z)\exp\left\{-\int \alpha(z)dz\right\}dz} \qquad (7.47)$$

Now with the measured values of Z_v using Equation (7.38) and by substituting the appropriate values of p_v and q_v in Equation (7.39), the water vapor content values may be obtained.

We are usually accustomed with different types of cloud along with their generation and formation mechanism. Among these types of clouds, we were concerned with clouds when the radiometric temperatures are little higher in the presence of cloud. These types of clouds sometimes extend well above the cirrus type and penetrate a few thousand feet into the stratosphere. The water content of cloud normally increases upward to a maximum in the vicinity of melting level. These types contain exceedingly high vertical velocities and in that case the presence of hail above the tropopause also is possible.

But, it is to be pointed out that such steady-state conditions with respect to water distribution in the cloud do not exist when updrafts equal or exceed the fall velocity of particle as rain. Such updrafts exist only locally for periods about 5–15 minutes in convective activity and hence can lead to high local concentration of water. However, the calculation of the columnar liquid water content of clouds from radiosonde measurements is based on the model proposed by Salonen et al. (1991).

The cloud detection is performed using the "critical humidity" function defined as

$$U_c = 1 - \sigma\alpha(1-\sigma)[1+\beta(\sigma-0.5)]$$ (7.48)

Here $\sigma = \frac{p}{p_0}$ which is ratio between the atmospheric pressure at the considered level and the pressure at the ground where $\alpha = 1.0$ and $\beta = \sqrt{3}$, as proposed by Salonen et al. (1991). The existence of cloud at the certain level was taken into consideration when the RH at a particular height is greater than the U_c, as mentioned in Equation (7.48). Again it is presumed that within the cloud layer, the water density, ρ_c, of any slice of the upper air sounding is a function of the air temperature, t (°C), and of the layer height, h (m). The existence of cloud at the certain level was taken into consideration when the RH at a particular height is greater than the U_c. The plot of the vertical profile of U_c (h) and RH for a particular day, April 10, 2009, over Brazil is presented in Figure 7.16. The cloud base height and the cloud top height are found by interpolation method as shown in the same figure.

Here h_1 is the base height of the cloud where the RH curve crosses the $U_c(h)$ curve for the first time. The cloud top h_2 is the point on the curve where the

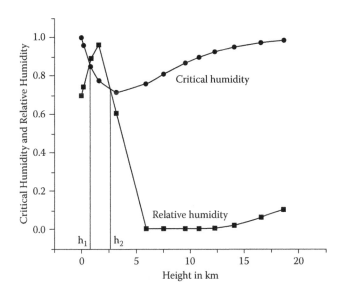

FIGURE 7.16
A simultaneous plot of critical humidity and relative humidity with height over Brazil (22° S).

curve $U_c(h)$ again crosses the RH curve. Again it is presumed that within the cloud layer the water density, ρ_c, of any slice of the upper air sounding is a function of the air temperature, t (°C), and of the layer height, h (m). It is given by

$$\rho_c(t,h) = \rho_0 \exp(ct) \left(\frac{h - h_b}{h_r} \right) \times p_w \tag{7.49}$$

where, $\rho_0 = 0.17(g/m^3)$; $c = 0.04$ (°C^{-1}) = temperature dependence factor; $h_r = 1500$ (m) and h_b = cloud base height (m).

And $p_w(t)$ is the liquid water fraction, approximated by

$$p_w(t) = 1, \quad \text{if } 0°C < t$$

$$p_w(t) = 1 + t/20, \quad \text{if } -20°C < t < 0°C$$

$$p_w(t) = 0, \quad \text{if } t < -20°C$$

The calculation of both cloud base and top heights may be worked out from the ground by linear interpolation. The columnar liquid water content, L, can be found by adding the contributions from all the layers within the clouds that contain water.

The total liquid water content is given by

$$L(kg/m^2) = \int_{h_1}^{h_2} \rho_L(h) \, dh \tag{7.50}$$

It is to be noted that over Brazil (22° south) the h_2 values were different for different days. But the maximum value of h_2 was found as 3 km. So it is conclusively decided that over Brazil, the cloud never was extended up to the stratosphere and question does not arise of formation of hail above the tropopause.

For clouds consisting entirely of small droplets, generally less than 0.01 cm, the Rayleigh approximation is valid for frequencies below 200 GHz, and it is possible to express the attenuation in terms of the total water content per unit volume. Thus the specific attenuation within a particular cloud can be written as

$$\alpha_c = K_L \rho_c \text{ dB/km} \tag{7.51}$$

where α_c is the specific attenuation (dB/km) within the cloud; K_L is the specific attenuation coefficient [(dB/km)/(g/m^3)]; and ρ_c is the liquid water density in the cloud (g/m^3).

A mathematical model based on Rayleigh scattering, which uses a double-Debye model for the dielectric permittivity, $\varepsilon(f)$, of water, can be used to calculate the value of, K_L for frequencies up to 1000 GHz:

$$K_L = \frac{0.819f}{\varepsilon''(1 + \eta^2)} \text{ (dB/km)/(g/m}^3) \tag{7.52}$$

where f is the frequency (GHz), and

$$\eta = \frac{2+\varepsilon'}{\varepsilon''}$$

The complex dielectric permittivity of water and corresponding relaxation frequencies are given by Equations (7.9) through (7.12).

This specific attenuation when integrated over the vertical path gives the total attenuation in decibels. The equation involved is

$$A_c(dB) = \int_0^\alpha \alpha(h)\,dh \qquad (7.53)$$

Brightness temperatures at different times of different months are made available from the radiometer data where from the total attenuation may be obtained using Equation (7.29). Now the measured total attenuation values (A_T) are used to obtain the values of Z_L using Equation (7.41). Liquid water content may be calculated using the Equation (7.42). The values of p_L and q_L may be obtained from the radiometric record for the desired period.

7.6.3.2 Use of Radiometric Data and Some Results over South America

We have already discussed that the present algorithm is developed for the true measurement of water vapor and cloud liquid. For this purpose, as a first step we have considered the radiometric measurement of the brightness temperature for the 22.234 and 30 GHz pair. The radiometric outputs in terms of brightness temperature in both the frequency channels were converted to attenuation (dB) by using Equation (7.29). This is presented in Figure 7.17. It shows that the attenuation at 22.234 GHz is approximately three times larger than that at 30 GHz. The month of April has been chosen because April is considered to be approximately the end of the monsoon season over South America (at 22° S). It is to be noted from Figure 7.17 that as time passes toward the end of April the attenuation values are going to be decreased. The measured values of attenuations at 22.234 and 30 GHz are then substituted in Equation (7.38) to find Z_v. Now, we have obtained the values of $p_v = 24.12534$ and $q_v = 0.91243$ for the month of April from the multiple regression analyses using radiosonde (refer to Equation 7.39) data. These along with measured vales of Z_v, integrated values of water vapor contents, V_m are measured.

A time series of the measured vapor has been shown in Figure 7.18. This has been compared in the same figure with those obtained by using radiosonde data. Figure 7.18 shows that the variation of water vapor content is in harmony with those obtained by using radiosonde data. It is also noted that the measured value always is slightly larger than the calculated values. This might be due to the fact that radiosonde measurement only is taking

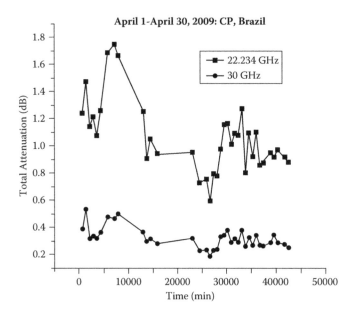

FIGURE 7.17
A time series of total radiometric attenuation (dB) at 22.234 and 30 GHz over Brazil during April 2009.

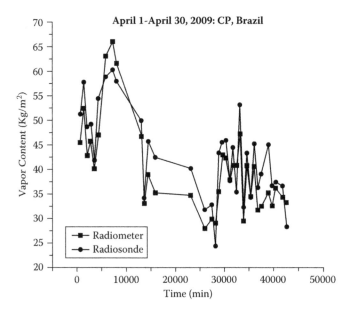

FIGURE 7.18
A time series of measured water vapor content (kg/m²) using dual frequency algorithm (22.234 and 30 GHz) and compared to corresponding radiosonde data.

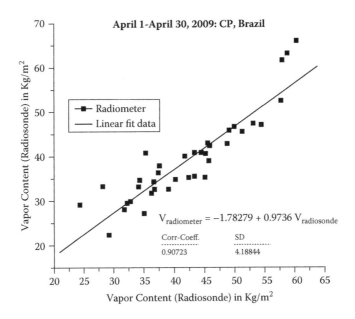

FIGURE 7.19
Relationship between the measured water vapor content (kg/m²) using 22.234 and 30 GHz and those obtained by using radiosonde data over Brazil.

care of water vapor, but on the other hand radiometric measurement counts all atmospheric constituents including water vapor. But still we do prefer radiometric measurement because of its continuity. Radiosonde data are only available twice daily. However, the radiometric measurement shows the variation of water vapor content from as low as 25 kg/m² up to a maximum 68 kg/m² over Brazil during April 2009. The best fit linear relation between the measured values of integrated water vapor content (kg/m²) with those obtained from the radiosonde data was found to be

$$V_{22.234\&30} = 0.93329\ V$$

where V is water vapor content obtained by using radiosonde data. In this case the correlation coefficient was 0.90723, standard deviation is 2.31 and percentage error is 4.88 and rms error is 1.689%. This is presented in Figure 7.19.

A time series of liquid water content has been plotted (refer to Figure 7.20). Here the values of $p_L = 1.767$ and $q_L = 0.0193$ are obtained by using the radiosonde data. It is to be noted here that the liquid content goes up to a maximum of 2.8 kg/m² during the month of April. This large value of liquid water content might be due to the presence of thick cloud over the antenna beam. The sudden fall of the graphical presentation is also observed at certain times. It might be due the presence of thin cloud overhead. The thick cloud may be dispersed with time due to wind activity that is happening for a very short period.

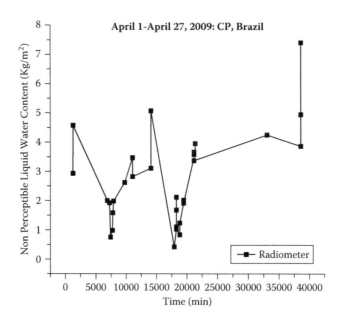

FIGURE 7.20
A time series of measured liquid water content (kg/m²) over Brazil, the frequency pair is 23.034 and 30 GHz.

But it is to be remembered that the choice of 22.234 GHz is not optimal because it is the line frequency with some collision broadening. This choice is pressure dependent and it is suggested to move slightly away from the resonance line and hence is the selection of a frequency pair consisting of 23.034 and 30 GHz.

Now following the same procedure, the relation between measured and calculated values of water vapor content has been studied. This is presented in Figure 7.21. The relation shows a linear fit bearing a relationship as

$$V_{23.034\&30} = 1.27 + 1.01\ V$$

where V is the calculated values of vapor content. This relation bears the correlation coefficient 0.93865; standard deviation is 1.35 and r.m.s error 3.42. Here it is to be noted that r.m.s error is less using 23.034 and 30 GHz pair instead of using 22.234 and 30 GHz pair.

A time series analysis of cloud liquid water content exploiting the 23.034 and 30 GHz pair has been presented in Figure 7.22. This experiment has been performed in the same climatic condition prevailed over the radiometer antenna, as it was done earlier in case of 22.234 and 30 GHz pair during April 2009. Figure 7.22 shows the maximum liquid water content over the antenna beam as 5.0 kg/m². This value is as good as double to that obtained earlier for the same event. So we can say that in case of the large liquid water bearing cloud, the 22.234 and 30 GHz pair is not capable enough to measure liquid water properly.

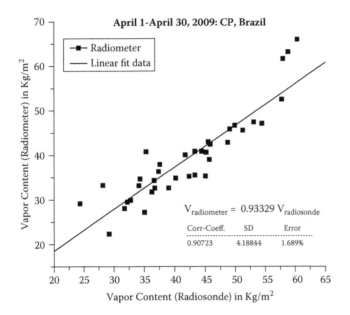

FIGURE 7.21
Relationship between the measured water vapor content (kg/m²) using 23.034 and 30 GHz and those obtained by using radiosonde data over Brazil.

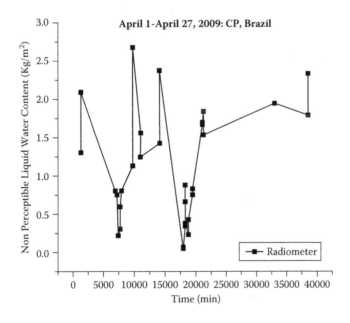

FIGURE 7.22
A time series of measured liquid water content (kg/m²) over Brazil.

The regression analysis between the estimated water vapor content by radiometric method and the brightness temperature from the radiometer at 22.234 GHz shows the best linear equations over Brazil as (Karmakar et al. 2010)

$$W = 478.451, T_b + 9574$$

A comparative study between the calculated (using the radiosonde data) values of water vapor content (y), and measured (using the radiometer) water vapor content (x) shows

$$Y = 0.4218\, x + 31.31$$

The same type of experiment was performed by Karmakar et al. (2001) at Kolkata (22° N) and found

$$V_m(22) = 1.13646\, V \text{ (percentage error 5\%)}$$

where V_m is the measured water vapor content by using 22.235 GHz only and V is the calculated vapor content by using radiosonde data. But the use of the dual frequency algorithm developed along with corresponding radiometric data analyses over Kolkata provide the relationship as (Karmakar et al. 2001)

$$V_{22.235\,\&31.4} = 1.0161\, V \text{ (percentage error 1.6062\%)}$$

The same experiment had been carried out for the measurement of vapor and liquid water by exploiting the frequency pair 23.834 and 30 GHz , keeping in mind that 23.834 GHz is a frequency which is considered to be independent of collisional broadening due to pressure. The corresponding result obtained is

$$V_{23.834\&30} = 3.195 + 0.876V$$

with an r.m.s error 2.64 and standard deviation is 1.51. So now, if we look at the tendency of minimizing the error percentage for using the dual frequency pair for the true measurement of water vapor in the ambient atmosphere, certainly the last choice is the optimum choice.

So it is worth noting that in estimating water vapor content by using the 22 GHz radiometer data only, the percentage error is 5%, whereas that for measurement of vapor using both 22.235 GHz and 31.4 GHz is 2% relative to the mean. However the liquid water content assumes values as low as 0.02 kg/m^2 and as high as 1.85 kg/m^2 over Kolkata.

So, it is conclusively decided to use the frequency in the vicinity of 30 GHz and the other one at around 22 GHz but not exactly the resonant line frequency occurring at 22.234 GHz for the true estimation of vapor content.

Besides these it was interestingly noted that over Brazil, the liquid water content is 1½ times larger than that obtained over Kolkata. It is also conclusively decided that in case of high liquid water content measurement, the frequency 22.234 GHz is not to be chosen instead it would be away from the resonant line. It should be somewhere a few gigahertz right or left from the resonant line to get a better accuracy.

Westwater and Guiraud (1980) applied another method for severe weather conditions in Lawton, Oklahoma, during March 29, 1979 to June 8, 1979. The frequencies used were 20.6 GHz and 31.65 GHz. Clouds were modeled by Decker et al. (1978). The percentage variations of the different parameters used, exclusive of the liquid attenuation, was less than 5%. However, the regression coefficients for the cloud were of the order of 25% to 30%. For water vapor sensing the departure of the regression coefficients for cloud from their average values are highly correlated with each other. The variation was only about 2%. However, it is to be noted that the frequency channel at 30 GHz is more sensitive than water vapor; the measured values would deviate more from the calculated values in presence of the cloud. For further details readers are referred to the work of Han et al. (1994) and in-depth analysis of ground based radiometric measurement of water vapor and liquid water (Han and Thomson 1994).

7.7 Effect of Water Vapor and Liquid Water on Microwave Spectra

It is commonly believed that the spectral behavior of atmospheric constituents in the microwave and millimeter wave band offers a good opportunity for the measurement of atmospheric constituents, such as water vapor and liquid water. The radiative properties of water vapor and liquid water are significantly different (Elgered 1993). We have seen already that because of this difference in spectral behavior, it is possible to separate water vapor and liquid water effects by measuring atmospheric attenuation at an additional frequency, away from the water vapor absorption peak occurring at 22.235 GHz. Karmakar et al. (2002) have found the water vapor attenuation coefficients (dB/km) over Kolkata in the range 1 to 40 GHz by using the millimeter wave propagation model (MPM) for the year 1996, as described by Liebe (1989). The integrated attenuation (dB) within the height limit 0 to 10 km was calculated using Romberg's integration method with the assumption that water vapor concentration above about 5 km has a negligible effect on the spectrum (Evans and Hagfors 1968). Figure 7.23 shows the attenuation coefficients in the range 1–40 GHz for a maximum limit of water vapor content 76 kg/m^2 and liquid water content varying from 0 to 2 kg/m^2 over Kolkata. It is clear from the figure that for a change of 2 kg/m^2, the first vapor maxima is approximately invariant in position, while the minima has been shifted from

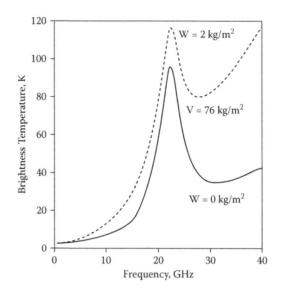

FIGURE 7.23
The brightness temperature as a function of frequency for a particular value of water vapor content (76 kg/m²) and liquid water content (2 kg/m² and 0 kg/m²).

31.448 GHz to 28.381 GHz for which a corresponding change in brightness temperature of about 46 K occurs, which is equivalent to 0.75 dB.

Figure 7.24 shows the spectrum for liquid water content set to zero varying the vapor content from 36 to 76 kg/m². This shows the minima lies at 31 GHz in both the cases, but a corresponding brightness temperature change of about 10 K occurs, for which the attenuation change is 0.11 dB.

So it appears that the effect of liquid water is more prominent than that of water vapor in determining the minima in the spectrum at a tropical location.

In fact, the variation of temperature and water vapor density are largely assumed to be the influencing factors in addition to liquid water, discussed earlier in determining the millimeter wave window frequencies over the chosen range of latitudes in between the two successive maxima occurring at 60 and 120 GHz. It is observed by Karmakar et al (2011) that between temperature and water vapor density, the later one is influencing mostly in determining the window frequency. It is also observed that the minima is occurring at 75 GHz through 94 GHz over the globe during the month January–February and 73 GHz through 85 GHz during the month July–August, depending on the latitudinal occupancy. It is observed that the large abundance of water vapor is mainly held responsible for shifting of minima towards the low value of frequencies. Hence, it is becoming most important to look at the climatological parameters in determining the window frequency at the place of choice. This may be well understood through the Table 7.8.

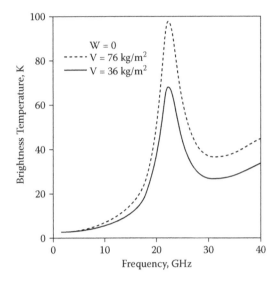

FIGURE 7.24
The brightness temperature as a function of frequency for a particular value of liquid water content and variable values of water vapor content (36 kg/m^2 and 76 kg/m^2).

7.8 Cloud Radar

The conventional weather radar is not capable enough to see the cloud drops. This is because the size of the cloud drops are usually around several micrometers to several ten micrometers, which is typically one hundredth of rain drop size and hence radar reflectivity factor is small. As the radar reflectivity is proportional to the sixth power of target size, much higher sensitivity is required for cloud observation. So, if we approach the higher frequency radar, it brings the higher sensitivity. Presently, the Ka band (near 35 GHz) and W band (near 95 GHz) are in use for cloud observation. Again we know that the back scattering cross-section in Rayleigh scattering increase in fourth power of radar wavelength. Now let us consider an increase in sensitivity with frequency by the Rayleigh effect with the assumption that other parameters such as antenna aperture and transmitter power are all identical. Hence, with this idea it can be shown that the 35 and 95 GHz radar has 22 dB and 39 dB higher sensitivity than conventional weather radar (10 GHz). Thus high frequency radar is suitable for cloud distribution measurement with high sensitivity in short range. In fact, this makes it possible due to recent developments of high power transmitters and low noise amplifiers in millimeter wavelength. In this respect the Communication Research Laboratory (CRL), Japan, has developed an airborne radar at 95 GHz (Okamoto 2001).

This radar provides the information about the various features of cloud, such as vertical profile, rain generating process, and atmospheric movement.

TABLE 7.8

Influence of Water Vapor and Temperature for Determining the Window Frequency over the Globe

Place	Country	Months	Temperature (K)	Water Vapor Density (g/m3)	Window Frequency (GHz)	Months	Temperature (K)	Water Vapor Density (g/m3)	Window Frequency (GHz)
Kolkata	India	July–Aug	303.45	24.39	73.69	Jan–Feb	299.15	12.91	77.42
Chongqing	China	July–Aug	302.72	20.04	74.48	Jan–Feb	282.81	6.57	81.50
Srinagar	India	July–Aug	291.22	13.71	75.98	Jan–Feb	277.58	4.86	82.46
Aldan	Russia	July–Aug	287.47	10.13	78.15	Jan–Feb	246.34	0.45	94.26
Lima Callao	Peru	July–Aug	289.60	10.59	78.25	Jan–Feb	293.85	15.54	76.00
Porto Alegre	Brazil	July–Aug	286.49	11.02	78.24	Jan–Feb	297.21	17.84	75.32
Paraparaumu	Newzeland	July–Aug	280.47	6.75	81.5	Jan–Feb	289.70	11.75	77.79
Comodoror	Argentina	July–Aug	278.31	3.79	85.35	Jan–Feb	284.29	6.97	81.64

This cloud structure would provide the information about the radiation budget of the atmosphere, which is pertinent to global warming research.

References

Adams, D. K. Rui, M. S., Fernandes, E. R., Maia, J. M., Sapucci, L. F., Machado, L. A. T., I. Vitorello., Monico, J. F. G., Kirk, L. H., Gutman, S.I., Filizola, N., and Bennett, R. A. A dense GNSS meteorological network for observing deep convection in the Amazon. Atmos. Sci. Let. Published online in Wiley Online Library (Wiley on line library.com) DOI: 10.1002/asl. 312, 2011.

Allnutt, J. E., 1976, Slant path attenuation and space diversity results using 11.6 GHz. Radiometer, *Proceedings of IEE*, 123, 1197–1200.

Alpert, P., Kufman, Y. J., Shay-El, Y., Tanre, D., Da Silva, A., Joseph, Y. H. Y., 1998, Dust forcing of climate inferred from correlation between dust data and model errors, *Nature*, 395, 367–370.

Aufm Kampe, H. J., Weick Mann, H. K., 1957, Physics of clouds, *Meteorol. Monographs*, 3, 182–201.

Bhattacharya, C. K., 1985, *Microwave radiometric studies of atmospheric water vapor and attenuation measurements* (PhD thesis), Benaras Hindu University, India.

Bohlander, R. A., Emery, R. J., Llewellyn-Jones, D. T., Gimmestad, G. G., Simpson, O. A., Gallagher, J. J., Perkowitz, S., 1980, Excess absorption by water vapor and comparison with theoretical dimer absorption. In *Atmospheric water vapor*, A. Depak, T. D. Wilkersor, L. H. Ruhnke, eds., 241–254, New York: Academic Press.

Braunt, D., 1947, *Physical and dynamical meteorology*, Cambridge: Cambridge University Press.

Bussy, H. E., 1950, Microwave attenuation estimated from rainfall and water vapor statistics, *Proceedings of the Institute of Radio Engineers*, 38, 781–785.

Buttorn, K. J., Wiltse, C. J., 1981, *Infrared and millimeter waves*, vol. 4, New York: Academic Press.

Charlson, R. Z., Schwartz, S. E., Hales, J. M., Cess, R. D., Coakley, J.A. Jr., Hansen, J. E., Hofman, D. J., 1992, Claimate forcing by anthropogenic aerosols, *Science*, 255, 423–430.

Decker, M. T., Westwater, E. R., Guiraud, F. O., 1978, Experimental evaluation of ground based remote sensing of atmospheric temperature and water vapor profiles, *Journal of Applied Meteorology*, 17, 1788–1795.

Donaldson, R. J. Jr., 1959, Analysis of severe convective storms observed by radar–II, *Journal of Meteorology*, 16, 281–286.

Donaldson, R. J. Jr., 1961, Radar reflectivity profiles in thunderstorms, *Journal of Meteorology*, 18, 292–305.

Elgered, G., 1993, Tropospheric radio path delay from ground based microwave radiometry. In *Atmospheric Remote Sensing by Microwave Radiometry*, M. A. Janssen (ed.), 215–258, New York: John Wiley & Sons.

Evans, J. V., Hagfors, T., 1968, *Radar astronomy*, New York: McGraw Hill.

Foster, D. S., 1961, *Aviation hail problems*, Technical Note, 37, World Meteorological Organisation.

Gibbins, C. J., 1986, Improved algorithm for the determination of specific attenuation at sea lovel by dry air and water vapor in the frequency range 1–350 GHz, *Radio Science*, 21, 949–954.

Gibbins, C. J., 1988, The effects of the atmosphere on radio wave propagation in the 50-70 GHz frequency band, *Journal of the Institution of Electronic and Radio Engineers*, 58, 6 (Supplement), S229–S240.

Han, Y., Snider, J. B., Westwater, E. R., Melfi, S. H., Ferrare, R. A., 1994, Observation of water vapor by ground based microwave radiometers and Raman lidar, *Journal of Geophysical Research*, 99, 695–702.

Han, Y., Thompson, D. W., 1994, Multichannel microwave radiometric observations at Saipan during the 1990 tropical cyclone motion experiment, *Journal of Atmospheric and Oceanographic Technology*, 11, 110–121.

Hansen, J., Sato, M., Ruedy, R., 1997, Radiative forcing and climate response. *Journal of Geophysical Research*, 102, 6831–6864.

Hansen, J., Sato, M., Ruedy, R., Lacis, A., Oinas, V., 2000, Global warming in the twenty-first century: An alternative scenario, *Proceedings of the National Academy of Science of the USA*, 97, 9875–9880.

Haywood, J. M., Ramaswamy, V., 1998, Global sensitivity studies of the direct radiative forcing due to anthropogenic sulfate and black carbon aerosols, *Journal of Geophysical Research*, 103, 6043–6058.

Haywood, J. M., Roberts, D. L., Slingo, A., Edwards, J. M., Shine, K. P., 1997, General circulation of the direct radiative forcing by anthropogenicsulfate and fossil-fuel soot aerosol, *Journal of Climate*, 10, 1562–1567.

Hess, L. T., 1959, *Introduction to theoretical meteorology*, New York: Heidy Holt.

Highwood, E. J., 2000, Effect of cloud inhomogeneity on direct radiative forcing due to aerosol, *Journal of Geophysical Research*, 105, 17843–17852.

Karmakar, P. K., 1989, *Studies of microwave and millimeter wave propagation* (PhD thesis), University of Calcutta, India.

Karmakar, P. K., Lahari, S., Maiti, M., Frederico, A. C., 2010, Some of the atmospheric influences on microwave propagation through atmosphere, *American Journal of Sc. and Ind. Research*, 1(2), 350–358.

Karmakar, P. K., Maiti, M., Calheiros., A. J. P., Angelis, C. F., Machado, L. A. T., and Da Costa, S. S., 2010, Ground based single frequency microwave radiometric measurement of water vapor, *International Journal of Remote Sensing*.

Karmakar, P. K., Maiti, M., Chattopadhyay, S., Rahaman, M., 2002, Effets of water vapor and liquid water on microwave absorption and their application, *Radio Science Bulletin*, 303, 32–36.

Karmakar, P. K., Maiti. M., Mondal, S., Angelis Carlos Frederico., 2011, Determination of window frequencyin the millimeter wave band in the range of 58° North through 45° south over the globe, accepted, Advances in Space Research, doi: 10.1016/ j.asr.2011.02.019.

Karmakar, P. K., Rahaman, M., Sen, A. K., 2001, Measurement of atmospheric water vapor content over tropical location by dual frequency microwave radiometry, *International Journal of Remote Sensing*, 22, 17, 3309–3322.

Karmakar, P. K., Sett, S. Maiti. M. Angelis. C. F., Machado, L. A., 2011, Radiometric estimation of water vapor content over Brazil, Accepted, Advances in Space Research, doi no. 10.1016/j.asr.2011.06.032.

Liebe, H. J., 1989, MPM—An atmospheric millimeter wave propagation model. *International Journal of Infrared and Millimeter Waves (USA)*, 10, 631–650.

Liebe, H. J., Manab, T., Hufford, G. A., 1989, Millimeter-wave attenuation and delay rates due to fog/cloud conditions, *IEEE Transactions on Antennas and Propagation*, 37, 1617–1623.

Liebe, H. J., 1985, An updated model for millimeter wave propagation in moist air, *Radio Science*, 20, 5, 1069–1089.

Miller, R. L., Tegen, I., 1998, Climate response to soil dust aerosols, *Journal of Climate*, 11, 3247–3267.

Mitra, A., Karmakar, P. K., Sen, A. K., 2000, A fresh consideration for evaluating mean atmospheric temperature, *Indian J. Physics* 74b, 5, 379–382

Moran, J. M., Rosen, B. R., 1981, Estimation of the propagation delay through the troposphere from microwave radiometer data, *Radio Science*, 16, 235–244.

Okamoto, K., ed., 2001, *Wave summit course: Global environment remote sensing*, Japan: IOS Press.

Pekney, N. J., Davidson, C. I. L., Hopke, P. K., 2006, Application of PSCF and CPF to PMF-modeled sources of Pm 2.5 in Pittsburg, *Aerosol Science and Technology*, 40, 952–953.

Resch, G. M., 1983, Another look at the optimum frequencies for a water vapor radiometer. *TDA Progress Report*, October–December 1883.

Ricchiazzi, P., Yang, S., Gautier, C. Andsoele, D., 1998, SBDART, A research and teaching software tool for plane-parallel radiative transfer in the Earth's atmosphere, *Bulletin of the American Meteorological Society*, 79, 2101–2114.

Salonen, E., 1991, New prediction method of cloud attenuation, *Electronics Letters*, 27, 12, 1106–1108.

Sen, A. K., Karmakar, P. K., Mitra, A., Devgupta, A. K., Dasgupta, M. K., Calla, O.P.N., Rana, S. S., 1990, Radiometric studies of clear air attenuation and atmospheric water vapor at 22.235 GHz, *Atmospheric Environment (Great Britain)*, 24A, 7, 1909–1913.

Sherwood, S. C., Roca, R., Weckwerth, T. M., Andronova, N. G., Tropospheric water vapor convection and climate: A critical review. Review of Geophysics 48, RG2001. DOI: 10.1029/2009RG000301, 2009.

Simpson, P. M., Brand, E. C., Wrench, C. L., 2002, Liquid water path algorithm development and accuracy—A report, Microwave radiometric measurement at Chilbolton, Radio Communication Research Unit, CLRC- Rutherford Appleton Laboratory, Chilton, Didcot, Oxon, 0X11OQX, U.K.

Tanre, D., Haywood, J., Pelon, J., Leon, J.F., Chatenet, B., Formenti, P., Francis, P., Goloub, P., Hoghwood, E. J., Myhre, G., 2003, Measurement and modelling of Saharan dust radiative impact: Overview of the Saharan dust experiment (SHADE), *Journal of Geophysical Research*, 108, 8574.

Torres, O., Bhartia, P. K., Herman, J. R., Sinyuk, A., Ginoux, P. P., Holben, B., 2002, A long term record of aerosol optical thickness from TOMS observations and comparison to AERONET measurements, *Journal of the Atmospheric Science*, 59, 398–413.

Tzanis, C., Varotsos, C. A., 2008, Tropospheric aerosol forcing of climate: A case study for the greater area of Greece, *International Journal of Remote Sensing*, 29, 9, 2507–2517.

Ulaby, F. T., Moore, R. K., Fung, A. K., 1986, *Microwave remote sensing: Active and passive*, vol. 1, Norwood, MA: Artech House.

Van de Hulst, H. C., 1957, *Light scattering by small particles*, New York: John Wiley & Sons.

Varotsos, C. A., Efstathiou, M. N., Kondratyev, K. Y., 2003, Long term variation in surface ozone and its precursors in Athens, Greece: A forecasting tool, *Environmental Science and Pollution Research*, 10, 19–23.

Varotsos, C. A., 2005, Power law correlations in column ozone over Antarctica, *International Journal of Remote Sensing*, 26, 3333–3342.

Varotsos, C. A., Kondratyev, K. Y., Cracknell, A. P., 2000, New evidence for ozone depletion over Athens, Greece. *International Journal of Remote Sensing*, 21, 2951–2955.

Varotsos, C. A., Kondratyev, K. Y., Katsikis, S., 1995, On the relationship between total ozone and solar ultraviolet radiation at St. Petersburg, Russia, *Geophysical Research Letters*, 22, 3481–3484.

Waters, J. W., 1976, Adsorption and emission by atmospheric gases. In *Methods of experimental physics*, M. L. Meeks, ed., New York: Academic Press.

Westwater, E. R., Guiraud, F. O., 1980, Ground-based microwave radiometric retrieval of precipitable water vapor in the presence of clouds with high liquid content, *Radio Science*, 13, 5, 947–957.

Appendix: Mean Atmospheric Temperature at Microwaves and Millimeter Waves in Clear Air Environment

The use of microwave and millimeter wave radiometric measurements of tropospheric temperature and integrated water vapor is well established by several authors. But the evaluation of the desired accuracy in radiometric measurement depends on several factors, such as the spatial and temporal humidity profile available from radiosonde data, which is not exact because the ascending balloon follows a random path for wind blow during telemetry. Besides this, the other source of error in brightness temperature measurement by a passive radiometer may creep in through the choice of mean atmospheric temperature, which is supposed to be the only parameter responsible in converting the antenna temperature to the attenuation values.

For example, let us assume a certain volume of the atmosphere that is considered to be an absorbing medium. This volume will attain a temperature T_m by absorbing the incident microwave energy from outside. It will reradiate isotropically. The extent of such absorption or emission of energy depends on fractional transmitivity (σ) of the atmospheric medium. Thus the radiated energy from the atmosphere is a noise that enhances the thermal noise temperature by an amount $(1 - \sigma)T_m$. This increase in thermal noise temperature is detected by the receiver as T_a, which is then given by

$$T_a = (1 - \sigma)T_m \text{ Kelvin}$$

Again by definition this attenuation A in decibels is related to σ by

$$A = 10 \log_{10}(1/\sigma)$$

This expression is evaluated to Equation 5.31. Thus in evaluating T_a from Equation 5.31, the assumption for the value of the fundamental parameter T_m seems to be a major source of error. As it is expected, T_m is a function of frequency and three basic meteorological parameters: atmospheric pressure, temperature, and dew point temperature. T_m has a significant role in very high frequency (VHF) and ultrahigh frequency (UHF) bands, but in the microwave and millimeter wave band the variation of T_m is appreciable. So to assess the variation of T_m we use the radiative transfer equation of

Chandrasekhar for nonscattering and nonrefractive atmosphere where the mean atmospheric temperature is given by

$$T_m = \frac{T_a}{\int \alpha_v \exp[\tau_v(0,z)]dz}$$

where T_a is the equivalent brightness temperature and is given by

$$T_a = \int_0^\infty \alpha_v(z)T(z)\left[\exp\left\{-\int_0^z \alpha_v(z)dz\right\}\right]dz$$

and $\tau_v(0,z) = \int_0^z \alpha_v(z)dz$, known as the zenith opacity.

Using the CCIR Recommendation (1990) developed from an updated propagation model of Liebe (1989), the zenith opacity (neper), specific attenuation (np/km), and brightness temperature may be calculated at the desired frequencies. For this purpose, Mitra et al. (2000) calculated the values of mean atmospheric temperature T_m at 22.235, 31.4, 53.75, 67.8, 76, 94, 118.75, 120.1, and 125 GHz over Kolkata (22° N) by using the radio-sonde data available from the India Meteorological Department. It was found there that in the nonmonsoon period, T_m takes the highest value of about 295.5K at 125 GHz and the lowest value of about 279K at 22.235 GHz. Moreover, it also shows that T_m remains more or less constant during the monsoon period (June–September) for all the selected frequencies. So it is evident that during the monsoon period when the water vapor content over Kolkata becomes maximum, the value of mean atmospheric temperature is nearly constant.

However, our specific goal is to estimate the appropriate value of T_m to find the attenuation at the desired frequency at the desired location from the measured radiometric value of brightness temperature. For this purpose, several attempts were made to express T_m in terms of readily available ground temperature (T_g). In this respect, Altshuler et al. (1968) gave an empirical relation as

$$T_m = 1.12T_g - 50 \text{ kelvin}$$

But the above relation sometimes overestimates the attenuation values for ground temperature when it lies around freezing point. Some of the workers prescribe the value of T_m for a particular frequency or frequency band. Brussaard (1985) prescribed T_m as 260 K for 11 GHz. Mitra et al. (2000) attempted to find a linear relation between the calculated values of T_m and T_g over Kolkata for the frequency band 22–140 GHz during monsoon and nonmonsoon months. They found

$$T_m = AT_g + B$$

TABLE A.1

Best Fit Linear Regression Coefficients for $T_m = AT_g + B$

Frequency (GHz)	Slope A (Kelvin/°C)	Intercept B (Kelvin)	Correlation Coefficient (r)
22.235	0.823	267.383	0.986
31.4	0.857	267.354	0.985
53.75	0.819	269.045	0.982
67.8	0.864	266.800	0.987
76.0	0.911	266.691	0.983
94.0	0.928	268.246	0.988
118.75	0.966	266.975	0.982
120.1	1.004	265.570	0.979
125.0	0.998	267.288	0.985

where A and B are the regression constants and are presented in Table A.1. Similarly, an attempt was made to relate dew point temperature (T_d) and surface water vapor density (ρ_0) with T_m. The empirical relations were

$$T_m = CT_d + D$$

$$T_m = E\,\rho_0 + F$$

where C, D and E, F are the regression constants, which are presented in Table A.2 and Table A.3. It was interesting that the correlations of T_m with T_d and ρ_0 are marginally better than that with surface temperature (T_g). However it is also to be mentioned that the value of regression constants make no sense unless the desired frequency and nature of profile of meteorological parameters are precisely defined.

TABLE A.2

Best Fit Linear Regression Coefficients for $T_m = CT_d + D$

Frequency (GHz)	Slope A (Kelvin/°C)	Intercept (Kelvin)	Correlation Coefficient (r)
22.235	0.778	270.05	0.985
31.4	0.816	270.00	0.991
53.75	0.779	271.59	0.988
67.8	0.828	269.36	0.990
76.0	0.870	269.46	0.992
94.0	0.880	271.20	0.990
118.75	0.923	269.89	0.992
120.1	0.963	268.53	0.992
125.0	0.950	270.39	0.991

TABLE A.3

Best Fit Linear Regression Coefficients for $T_m = E\rho_0 + F$

Frequency (GHz)	Slope A (Kelvin/°C)	Intercept B (Kelvin)	Correlation Coefficient (r)
22.235	0.748	272.15	0.979
31.4	0.786	272.18	0.987
53.75	0.749	273.70	0.982
67.8	0.799	271.55	0.988
76.0	0.838	271.77	0.988
94.0	0.846	273.57	0.985
118.75	0.889	272.35	0.989
120.1	0.929	271.07	0.989
125.0	0.915	272.92	0.987

According to Ulaby et al. (1986), T_m is usually estimated from the surface temperature. Based on 24 radiosonde profiles, Wu (1979) developed a simple relation $T_m = aT_g$. It is found there that $a \cong 0.95$ for frequencies between 20 and 24.5 GHz and $a \cong 0.94$ for 31.4 GHz. But Mitra et al. found $a \cong 0.86$ for 31.4 GHz with a correlation of 0.98.

So it is suggested to find the appropriate value for T_m at a particular place of choice. Moreover it is prescribed to use either the dew point temperature or the surface water vapor density as the parameter for evaluating the mean atmospheric temperature empirically.

References

Altshuler, E. E., Falcone, V. J., Wulfsberg, K. N., 1968, Atmospheric effects on propagation at millimeter wavelengths, *IEEE Spectrum*, 5, 83–90.

Brussaard, G., 1985, Radiometry: A useful prediction tool? Noordwijk: ESA Scientific and Technical Publication Branch, SP-1071.

CCIR, 1990, Effects of multipath on digital transmission over links in the maritime mobile-satellite service, Report 762-2 Annex to vol. VIII, Geneva: ITU, 386–401.

Liebe, H. J., 1989, MPM: An atmospheric millimeter wave propagation model, *International Journal of Infrared and Millimeter Waves*, 10, 631–650.

Mitra, A., Karmakar, P. K., Sen, A. K., 2000, A fresh consideration for evaluating mean atmospheric temperature, *Indian Journal of Physics*, 74b, 5, 379–382.

Ulaby, F. T., Moore, R. K., Fung, A. K., 1986, *Microwave remote sensing: Active and passive*, vol. 3, *From theory to application*, Norwood, MA: Artech House.

Wu, S. C., 1979, Optimum frequencies of a passive radiometer for tropospheric path length correction, *IEEE Trans Antenna and Propagation*, AP-27, 233–239.

Index

Printed and bound by CPI Group (UK) Ltd, Croydon, CR0 4YY

18/10/2024

01776261-0002